ZIML Math Competition Book

Varsity Division 2017-2018

Areteem Institute

ZIML Math Competition Book Varsity Division 2017-18

Edited by   Kevin Wang
              John Lensmire
              David Reynoso
              Kelly Ren

ISBN: 1-944863-31-1
ISBN-13: 978-1-944863-31-9
First printing, October 2018.

Math Challenge III Geometry
Math Challenge I-B Counting and Probability
Math Challenge II-A Combinatorics
Math Challenge I-B Number Theory
Math Challenge II-A Number Theory

COMING SOON FROM ARETEEM PRESS

Fun Math Problem Solving For Elementary School Vol. 2 (and Solutions Manual)
Counting & Probability for Middle School (and Solutions Manual) - From Common Core to Math Competitions
Number Theory Problem Solving for Middle School (and Solutions Manual) - From Common Core to Math Competitions
Other volumes in the **Math Challenge Curriculum Textbooks Series**

The books are available in paperback and eBook formats (including Kindle and other formats).
To order the books, visit https://areteem.org/bookstore.

Contents

Introduction

Each month during the school year, Areteem Institute hosts the online Zoom International Math League (ZIML) competitions. Students can compete in one of five divisions based on their age and mathematical level (details shown on Page 9).

This book contains the problems, answers, and full solutions from the nine ZIML Varsity Division Competitions held during the 2017-2018 School Year. It is divided into three parts:

1. The complete Varsity Division ZIML Competitions (20 questions per competition) from October 2017 to June 2018.
2. The solutions for each of the competitions, including detailed work and helpful tricks.
3. An appendix including the topics and knowledge points covered for Varsity Division, a glossary including common mathematical terms, and answer keys for each of the competitions so students can easily check their work.

The questions found on the ZIML competitions are meant to test your problem solving skills and train you to apply the knowledge you know to many different applications. We hope you enjoy the problems!

About Zoom International Math League

The Zoom International Math League (ZIML) has a simple goal: provide a platform for students to build and share their passion for math and other STEM fields with students from around the globe. Started in 2008 as the Southern California Mathematical Olympiad, ZIML has a rich history of past participants who have advanced to top tier colleges and prestigious math competitions, including American Math Competitions, MATHCOUNTS, and the International Math Olympaid.

The ZIML Core Online Programs, most available with a free account at ziml.areteem.org, include:

- **Daily Magic Spells:** Provides a problem a day (Monday through Friday) for students to practice, with full solutions available the next day.
- **Weekly Brain Potions:** Provides one problem per week posted in the online discussion forum at ziml.areteem.org. Usually the problem does not have a simple answer, and students can join the discussion to share their thoughts regarding the scenarios described in the problem, explore the math concepts behind the problem, give solutions, and also ask further questions.
- **Monthly Contests:** The ZIML Monthly Contests are held the first weekend of each month during the school year (October through June). Students can compete in one of 5 divisions to test their knowledge and determine their strengths and weaknesses, with winners announced after the competition.
- **Math Competition Practice:** The Practice page contains sample ZIML contests and an archive of AMC-series tests for online practice. The practices simulate the real contest environment with time-limits of the contests automatically controlled by the server.
- **Online Discussion Forum:** The Online Discussion Forum

is open for any comments and questions. Other discussions, such as hard Daily Magic Spells or the Weekly Brain Potions are also posted here.

These programs encourage students to participate consistently, so they can track their progress and improvement each year.

In addition to the online programs, ZIML also hosts onsite Local Tournaments and Workshops in various locations in the United States. Each summer, there are onsite ZIML Competitions at held at Areteem Summer Programs, including the National ZIML Convention, which is a two day convention with one day of workshops and one day of competition.

ZIML Monthly Contests are organized into five divisions ranging from upper elementary school to advanced material based on high school math.

- **Varsity:** This is the top division. It covers material on the level of the last 10 questions on the AMC 12 and AIME level. This division is open to all age levels.
- **Junior Varsity:** This is the second highest competition division. It covers material at the AMC 10/12 level and State and National MathCounts level. This division is open to all age levels.
- **Division H:** This division focuses on material from a standard high school curriculum. It covers topics up to and including pre-calculus. This division will serve as excellent practice for students preparing for the math portions of the SAT or ACT. This division is open to all age levels.
- **Division M:** This division focuses on problem solving using math concepts from a standard middle school math curriculum. It covers material at the level of AMC 8 and School or Chapter MathCounts. This division is open to all students who have not started grade 9.

- **Division E:** This division focuses on advanced problem solving with mathematical concepts from upper elementary school. It covers material at a level comparable to MOEMS Division E. This division is open to all students who have not started grade 6.

This problem book features the Varsity Division Contests. For a detailed list of topics covered for Varsity Division see p.189 in the Appendix.

About Areteem Institute

Areteem Institute is an educational institution that develops and provides in-depth and advanced math and science programs for K-12 (Elementary School, Middle School, and High School) students and teachers. Areteem programs are accredited supplementary programs by the Western Association of Schools and Colleges (WASC). Students may attend the Areteem Institute through these options:

- Live and real-time face-to-face online classes with audio, video, interactive online whiteboard, and text chatting capabilities;
- Self-paced classes by watching the recordings of the live classes;
- Short video courses for trending math, science, technology, engineering, English, and social studies topics;
- Summer Intensive Camps on prestigious university campuses and Winter Boot Camps;
- Practice with selected daily problems for free, and monthly ZIML competitions at ziml.areteem.org.

The Areteem courses are designed and developed by educational experts and industry professionals to bring real world applications into STEM education. The programs are ideal for students who wish to build their mathematical strength in order to excel academically and eventually win in Math Competitions (AMC, AIME, USAMO, IMO, ARML, MathCounts, Math Olympiad, ZIML, and other math leagues and tournaments, etc.), Science Fairs (County Science Fairs, State Science Fairs, national programs like Intel Science and Engineering Fair, etc.) and Science Olympiad, or purely want to enrich their academic lives by taking more challenges and developing outstanding analytical, logical thinking and creative problem solving skills.

Since 2004 Areteem Institute has been teaching with methodology that is highly promoted by the new Common Core State Standards: stressing the conceptual level understanding of the math concepts, problem solving techniques, and solving problems with real world applications. With the guidance from experienced and passionate professors, students are motivated to explore concepts deeper by identifying an interesting problem, researching it, analyzing it, and using a critical thinking approach to come up with multiple solutions.

Thousands of math students who have been trained at Areteem achieved top honors and earned top awards in major national and international math competitions, including Gold Medalists in the International Math Olympiad (IMO), top winners and qualifiers at the USA Math Olympiad (USAMO/JMO), and AIME, top winners at the Zoom International Math League (ZIML), and top winners at the MathCounts National. Many Areteem Alumni have graduated from high school and gone on to enter their dream colleges such as MIT, Cal Tech, Harvard, Stanford, Yale, Princeton, U Penn, Harvey Mudd College, UC Berkeley, UCLA, etc. Those who have graduated from colleges are now playing important roles in their fields of endeavor.

Further information about Areteem Institute, as well as updates and errata of this book, can be found online at `http://www.areteem.org`.

Acknowledgments

This book contains the Online ZIML Varsity Division Problems from the 2017-18 school year. These problems were created and compiled by the staff of Areteem Institute. These problems were inspired by questions from the Areteem Math Challenge Courses, past questions on the ACT/SAT/GRE, past math competitions, math textbooks, and countless other resources and people encountered by the Areteem Curriculum Department in their life devoted to math. We thank all these sources for growing and nurturing our passion for math.

The Areteem staff, including John Lensmire, David Reynoso, Kevin Wang, and Kelly Ren, are the main contributors who compiled, edited, and reviewed this book.

Lastly, thanks to all the students who have participated and continue to participate in the Zoom International Math League. Your dedication to the Daily Magic Spells and Monthly Contests makes all of this possible, and we hope you continue to enjoy ZIML for years to come!

1. ZIML Contests

This part of the book contains the Varsity Division ZIML Contests from the 2017-18 School Year. There were nine monthly competitions, held on the dates found below:

- October 6-8
- November 3-5
- December 1-3
- January 5-7
- February 2-4
- March 2-4
- April 6-8
- May 4-6
- June 1-3

1.1 ZIML October 2017 Varsity Division

Below are the 20 Problems from the Varsity Division ZIML Competition held in October 2017.

The answer key is available on p.203 in the Appendix.

Full solutions to these questions are available starting on p.74.

Problem 1

Consider lines $\ell_1 : y = x - 2, \ell_2 : y = x + 2, \ell_3 : y = 6 - x$. Consider circles of the form $(x - H)^2 + (y - K)^2 = S$ that are tangent to all 3 of ℓ_1, ℓ_2, ℓ_3. What is the smallest possible value of $H + K + S$?

Problem 2

Find the remainder when $(57^{37} + 46)^{26}$ is divided by 50.

Problem 3

Using all the digits $1, 2, 3, 4, 5$, how many 5-digit numbers can be formed so that the hundreds digit is NOT 3, and the 5-digit number is greater than 20000.

Problem 4

Suppose that a and b are real numbers and that $2x^2 + ax + b = 0$ has real solutions x_1 and x_2 with $x_1^2 + x_2^2 = 1$. Pairs (a, b) exist for all $K \leq b \leq L$ for real numbers K, L. What is $L - K$, rounded to the nearest tenth if necessary.

Problem 5

Given that

$$\binom{9}{0} + \frac{1}{2}\binom{9}{1} + \frac{1}{3}\binom{9}{2} + \frac{1}{4}\binom{9}{3} + \cdots + \frac{1}{10}\binom{9}{9} = \frac{a}{b}$$

where a and b are positive integers and $\gcd(a,b) = 1$. Find $a + b$.

Problem 6

Suppose you have 8 identical balls. How many ways are there to put them in 5 numbered boxes so that at least one of the boxes gets at least 3 balls?

Problem 7

In quadrilateral $ABCD$, $\angle B = \angle D = 90°, \angle A = 60°$. Also given that $AB = 4, AD = 5$. Find BC/CD.

Problem 8

Find the sum of all real roots of $\sqrt[3]{5 - x} + \sqrt{x - 4} = 1$.

Problem 9

Find all ordered pairs (a, b) such that $a^2 + b^2 = 629$, and the equation $x^2 + ax + 1 = b$ has two positive integer roots. What is $|a| + |b|$?

Problem 10
In triangle ABC with $\angle A = 120°$. Bisectors $\overline{AD}$, $\overline{BE}$ and $\overline{CF}$ meet at point O. Find the measure of $\angle DFO$ in degrees.

Problem 11
Suppose 3 numbers are chosen from the set $\{1, 2, \ldots, 9\}$. In how many ways can this be done such that the chosen subset has at least one pair of consecutive numbers?

Problem 12
Let a and b be positive integers such that $a + b = 120$, $\gcd(a, b) = 24$, and $\text{lcm}(a, b) = 144$. Given that $ar < b$, find the value of $100a + b$.

Problem 13
In a tetrahedron, consider the 4 vertices and midpoints of 6 edges, in total 10 points. Among them, how many groups of 4 points are NOT coplanar? (In other words, the 4 points are not on the same plane)

Problem 14
Let $M = 1 + 2 + 3 + \cdots + 2016$. Find the remainder when 2016! is divided by M.

Problem 15

Assume $z^7 = 1$ and $z \neq 1$. Evaluate the following:

$$z^3 + \frac{1}{z^3} + z^6 + \frac{1}{z^6} + z^9 + \frac{1}{z^9}.$$

Problem 16

The positive integers a, b, c and d are all divisible by the positive integer $ab - cd$. Find the sum of all possible values of $ab - cd$.

Problem 17

Let O be the intersection of the diagonals of convex quadrilateral $ABCD$. Given that $[ABC] = 5, [ACD] = 10$, and $[ABD] = 6$, find $[ABO]$. Round your answer to the nearest tenth if necessary. (Here, for example, $[ABC]$ denotes the area of $\triangle ABC$.)

Problem 18

Let P be an interior point of square $ABCD$, $PA = 5$, $PD = 8$, and $PC = 13$. Find the area of square $ABCD$.

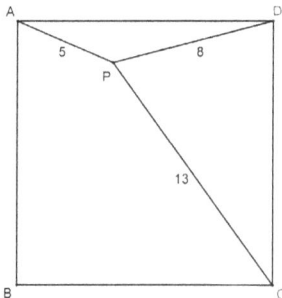

Problem 19

Let x_1 be a root of the equation $\log_3 x + x - 3 = 0$, and x_2 be a root of the equation $3^x + x - 3 = 0$, find the value of $x_1 + x_2$.

Problem 20

For each positive odd integer $n < 10000$, denote as $f(n)$ the number formed by the last four digits of n^9. The set A consists of all odd integers $n < 10000$ such that $f(n) > n$ and set B consists of all odd integers such that $f(n) < n$. Let $|A|$ and $|B|$ denote the numbers of elements in sets A and B respectively, find $|A| - |B|$.

1.2   ZIML November 2017 Varsity Division

Below are the 20 Problems from the Varsity Division ZIML
Competition held in November 2017.
The answer key is available on p.204 in the Appendix.
Full solutions to these questions are available starting on p.85.

Problem 1
Suppose you place 9 different rings on the 3 mid fingers of your
left hand (that is, not on your thumb or your pinky finger). In
how many different ways can this be done if no finger has more
than 3 rings?

Problem 2
Let x be the real root of the following equation:

$$\sqrt[3]{(x+1)^2} + 2\sqrt[3]{(x-1)^2} = 3\sqrt[3]{x^2-1}.$$

Suppose $x = \dfrac{m}{n}$, where m and n are positive integers and
$\gcd(m,n) = 1$. Find the value of $m+n$.

Problem 3
Assume $m > 1$. Given that four numbers, 2836, 4582, 5164, and
6522 have the same remainder $r > 0$ when divided by m. Find
the sum of all possible values of r.

Problem 4
In rectangle $ABCD$, $AB = 5$ and $BC = 7$. Fold the rectangle and flatten it so that A meets C. The area of the overlap region (after the folding) is $\dfrac{P}{Q}$ for positive integers P, Q with $\gcd(P, Q) = 1$. What is $P - Q$?

Problem 5
Suppose you have 3 real numbrs a, b, c. $a = 2$ and b is randomly chosen from the interval $[0, 3]$ while c is randomly chosen from the interval $[0, 4]$. The probability that you can make a triangle with side lengths a, b, c is $L\%$. What is L? Round your answer to the nearest tenth if necessary.

Problem 6
Find the sum of all even positive divisors of 10000.

Problem 7
Suppose equilateral triangle ABC is inscribed in a circle. Let P be on minor arc $\overarc{AC}$ such that $AP = 3$ and $BP = 5$ (distances of the line segments). Find the distance PC.

Problem 8
Let a, b, c, d be the roots of $x^4 - 5x^2 + 3x + 1$. Evaluate

$$a^4 + b^4 + c^4 + d^4.$$

Problem 9

71 basketball tickets are to be distributed to 3 groups of students, where any group cannot get more tickets than the total of the other two groups. In how many ways can this be done?

Problem 10

Let G be the centroid of $\triangle ABC$, and $AG = 3, BG = 4, CG = 5$, find $[ABC]$. Here $[ABC]$ denotes the area of $\triangle ABC$.

Problem 11

Find the last two digits of $7^{7^{.^{.^{.^7}}}}$ where there are seven total 7s.

Problem 12

How many quadruples (a, b, c, d) of non-negative integers satisfy the inequality $a + b + c + d \leq 23$?

Problem 13

In $\triangle ABC$, $BC = 14, AC = 9$, and $AB = 13$, the incircle is tangent to sides $\overline{BC}, \overline{AC}$, and $\overline{AB}$ at D, E, and F respectively. Calculate $AF^2 + BD^2 + CE^2$.

Problem 14

Evaluate $\log_{10}\left(\sqrt{3 + \sqrt{5}} + \sqrt{3 - \sqrt{5}}\right)$. Round your answer to the nearest tenth if necessary.

Problem 15

Let $A_1 A_2 \cdots A_n$ be a regular n-gon, where $n \geq 4$. Given that $\dfrac{1}{A_1 A_2} = \dfrac{1}{A_1 A_3} + \dfrac{1}{A_1 A_4}$, find n.

Problem 16

Find the maximum positive integer n such that $\dfrac{(n-3)(n+2)}{2n-1}$ is an integer.

Problem 17

The smallest positive real number x satisfying

$$\frac{4x^2}{(1 - \sqrt{1+2x})^2} \geq 2x + 9$$

can be written as $\dfrac{P}{Q}$ for positive integers P, Q with $\gcd(P, Q) = 1$. What is $P + Q$?

Problem 18

Let

$$S = 1 + \frac{1}{\sqrt{3}} + \frac{1}{\sqrt{5}} + \cdots + \frac{1}{\sqrt{289}}.$$

Find the value of $\lfloor S \rfloor$. Recall $\lfloor x \rfloor$ is the greatest integer not exceeding x.

Problem 19

How many integral solutions to the equation

$$a + b + c + d = 100,$$

are there given the following constraints:

$$1 \leq a \leq 10, \ b \geq 0, \ c \geq 2, 20 \leq d \leq 30?$$

Problem 20

Assume $a, b, c > 0$ and let

$$f(a,b,c) = \frac{a}{b+c} + \frac{4b}{c+a} + \frac{5c}{a+b}.$$

If L is the smallest integer such that $f(a,b,c) \leq L$ for some a, b, c, what is L?

1.3 ZIML December 2017 Varsity Division

Below are the 20 Problems from the Varsity Division ZIML Competition held in December 2017.
The answer key is available on p.205 in the Appendix.
Full solutions to these questions are available starting on p.97.

Problem 1
As shown in the diagram, three squares are lined up. The side lengths of the two smaller squares are 12 and 8 respectively. Find the area of $\triangle ABC$.

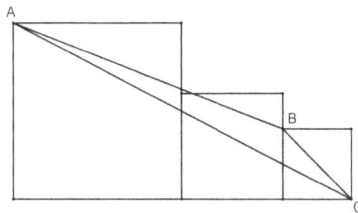

Problem 2
Four teachers are assigned to three (different) classes to teach, where each class should have at least one teacher. If all the teachers are assigned to a class, how many different assignments can be made?

Problem 3
The $x^2 = \lfloor 3x \rfloor$ has integer solutions of $x = 0$ and $x = 3$. There are two other solutions which are irrational. The smaller irrational solution can be written as $\sqrt{R}$ for an integer R. What is R? (Here $\lfloor x \rfloor$ represents the greatest integer not exceeding x)

Problem 4

Carrie writes the numbers $1, 2, 3, \ldots, 500$ (in order) on the board. Starting from the front, she erases one number, keeps one number, erases one number, etc. until she works through the entire list once. She then starts over at the beginning of the (now smaller) list, erasing the first number, keeping the second, etc. This continues (each time starting over at the beginning of the list) until there is only one number remaining on the board. What is this number?

Problem 5

Randomly place 5 distinct balls into 10 distinct boxes. The probability that the last box contains exactly 3 balls can be written as $\dfrac{P}{Q}$ for positive integers P, Q with $\gcd(P, Q) = 1$. What is $P + Q$?

Problem 6

Solve the equation for x:

$$8^{\log_6(x^2 - 8x + 16)} = 4^{\log_6 8}.$$

What is the (positive) difference between the solutions? Round your answer to the nearest integer if necessary.

Problem 7

A positive integer is written on each face of a cube. Each vertex is then assigned the product of the numbers written on the three faces intersecting the vertex, with a partial example shown below.

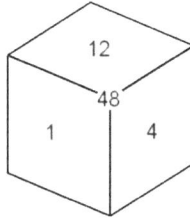

The sum of the numbers assigned to all the vertices is equal to 2015. Find the sum of the numbers written on the faces of the cube.

Problem 8

Consider the set $S = \{1, 2, 3, 4, 5, 6, 7\}$. Choose non-empty subsets A and B from S such that the smallest number in A is greater than the largest number in B. In how many ways can these subsets be chosen?

Problem 9

The equation $|x^2 - 5x| = A$ has exactly two distinct real roots. For how many integers $A \leq 100$ is this true?

Problem 10

Three mutually-tangent spheres, each with radius 10 inch, rest on the ground. These spheres are fixed in place and a fourth identical sphere is added (on the top) so that the fourth sphere is tangent to all the other spheres. If the total height of the shape formed by the four spheres is H, what is $\lfloor H \rfloor$? Here $\lfloor x \rfloor$ is the greatest integer not exceeding x.

Problem 11

Let n be a single digit (at most 9). Use the digits $1, 2, \ldots, n$ to form n-digit numbers with no repeating digits, where 2 cannot be adjacent to 1 or 3. Assume there are 2400 such numbers in total. Find the value of n.

Problem 12

Given $\triangle ABC$ with $\cos B = \dfrac{24}{25}, \cos C = \dfrac{12}{13}, \sin A = \dfrac{P}{Q}$ where $\gcd(P, Q) = 1$. What is $Q - P$?

Problem 13

What is the sum of all the integer solutions to

$$(x+1)(x+3)(x+5)(x+7) + 15 = 0?$$

Problem 14

Consider triples of positive integers (A, B, C) such that

$$A \equiv B \pmod{C}, \quad B \equiv C \pmod{A}, \quad C \equiv A \pmod{B}.$$

How many such triples are there with $1 \leq A \leq B \leq C \leq 5$?

Problem 15

Find the smallest prime factor of $29! - 1$.

Problem 16

Assume $x > 2, y > 2$, and let

$$a = \lfloor \log_2 x \rfloor, \quad b = \{\log_2 x\}, \quad c = \lfloor \log_2 y \rfloor, \quad d = \{\log_2 y\}.$$

Also given that $|1 - a| + \sqrt{c - 4} = 1$ and $b + d = 1$, find the value of xy. (Here $\lfloor r \rfloor$ represents the greatest integer not exceeding real number r, and $\{r\} = r - \lfloor r \rfloor$.)

Problem 17

There are 30 marbles in a bag, labeled 1 through 30. Suppose the weight of marble with label n is equal to $n^2 - 20n + 103$ grams. Select the marbles from the bag with equal probability without considering the weights. If two marbles are selected at once, the probability the two marbles have the same weight can be written as a simplified fraction $\dfrac{P}{Q}$. What is $P + Q$?

Problem 18

In a right triangle, the sum of the lengths of two legs is 63, and the altitude on the hypotenuse is 16. Find the length of the hypotenuse.

Problem 19

Given an angle $\angle POQ = 40°$. Let A be a point on $\overrightarrow{OQ}$ and B be a point on $\overrightarrow{OP}$, where $OA = 3, OB = 4$. Pick an arbitrary point A_1 on $\overline{OB}$ and another arbitrary point B_1 on $\overline{AQ}$. Let

$$k = AA_1 + A_1B_1 + B_1B.$$

The minimum value of k can be expressed as $\sqrt{N}$ for an integer N. What is N?

Problem 20

Consider the remainder when $x^{1234} + x^{2341} + x^{3412} + x^{4123}$ is divided by $x^4 + x^3 + x^2 + x + 1$. What is the constant term of this remainder?

1.4   ZIML January 2018 Varsity Division

Below are the 20 Problems from the Varsity Division ZIML Competition held in January 2018.

The answer key is available on p.206 in the Appendix.

Full solutions to these questions are available starting on p.107.

Problem 1
Find the last two digits of 2018^{2018}.

Problem 2
Real numbers x, y satisfy

$$\begin{cases} \sqrt{12 - y} = \sqrt{x} + \sqrt{x - y}, \\ \sqrt{4 - y} = \sqrt{x} - \sqrt{x - y}. \end{cases}$$

What is $x^2 + y^2$?

Problem 3
Separate 12 students into 6 pairs. In how many ways can this be done?

Problem 4
In quadrilateral $ABCD$, $\angle A = 60°$, $\angle B = \angle D = 90°$, $AB = 11\sqrt{3}$, $BC = 15$. Find the length of the diagonal $\overline{BD}$.

Problem 5
What is the sum of all positive integers less than 100 and relative prime to 100?

Problem 6
In the expansion of $(x+y+z+w)^{10}$, what is the coefficient of the term $x^4 y^3 z^2 w$?

Problem 7
Assume $m > 1$. Given that four numbers, 2836, 4582, 5164, and 6522 have the same remainder $r > 0$ when divided by m. Find maximum possible value of r.

Problem 8
Let ω be a complex number satisfying $\omega^2 + \omega + 1 = 0$. Find the value of
$$(1 - \omega)(1 - \omega^2)(1 - \omega^4)(1 - \omega^8).$$

Problem 9
In $\triangle ABC$, $\angle A = 60°$. The inscribed circle touches side $\overline{BC}$ at D. Given that $BD = \sqrt{3}, DC = 2$, find the area of $\triangle ABC$.

Problem 10

A debate club consists of 8 boys and 7 girls. In a debate competition, a team of 2 boys and 2 girls from this club is to be selected and seated in 4 chairs on the stage. In how many ways can this be done?

Problem 11

How many positive integers not exceeding 1000 do not have any of 2, 3, or 5 as factors?

Problem 12

Evaluate
$$\csc 10° \csc 30° \csc 50° \csc 70°.$$

Problem 13

Find the maximum possible real number x that satisfy the equation:
$$\sqrt{x+3-4\sqrt{x-1}}+\sqrt{x+8-6\sqrt{x-1}}=1.$$

Problem 14

Let $ABCDEFGH$ be an octagon inscribed in a circle. Four of the sides of the rectangle $AB = BC = CD = DE = 1$, and the remaining four sides $DE = EF = FG = GH = \sqrt{2}$, find the area of the octagon.

Problem 15

Find the sum of all positive integers n such that $\left\lfloor \dfrac{n^2}{4} \right\rfloor$ is prime. Here $\lfloor x \rfloor$ represents the greatest integer not exceeding x.

Problem 16

From the set of positive integers $\{1, 2, 3, \ldots, 15\}$, select 3 of them, so that the sum of the 3 selected integers is divisible by 3. In how many ways can this be done?

Problem 17

In $\triangle ABC$, D is a point on $\overline{BC}$ such that $\dfrac{BD}{DC} = \dfrac{1}{3}$. Let E be the midpoint of $\overline{AC}$, O be the intersection of $\overline{AD}$ and $\overline{BE}$, F be the intersection of $\overline{CO}$ and $\overline{AB}$. Find the ratio between the area of $\triangle ABC$ and the area of quadrilateral $BDOF$.

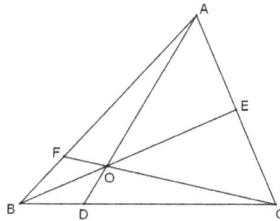

Problem 18
Let x and y be real numbers satisfying

$$\begin{cases} x+y = -20, \\ \sqrt[3]{x-1}+\sqrt[3]{y+2} = -1. \end{cases}$$

Find the sum of all possible values of xy.

Problem 19
Let x,y,z be digits. The base-10 representation of the positive integer n is $\overline{13xy45z}$. Given that $792 \mid n$, find the value of $100x+10y+z$.

Problem 20
Let n be a positive integer, and let real numbers $x_1,x_2,\ldots,x_n$ satisfy

$$\frac{x_1}{x_1^2+1}=\cdots=\frac{x_n}{x_n^2+1},$$
$$x_1+\cdots+x_n+\frac{1}{x_1}+\cdots+\frac{1}{x_n}=\frac{10}{3}.$$

Define $y_i = \log_3 x_i$ for $1 \le i \le n$. Find the value of $|y_1|+\cdots+|y_n|$.

1.5 ZIML February 2018 Varsity Division

Below are the 20 Problems from the Varsity Division ZIML Competition held in February 2018.
The answer key is available on p.207 in the Appendix.
Full solutions to these questions are available starting on p.117.

Problem 1

Consider the set of all right triangles (up to congruence) that have integer side lengths and have one side measuring 29. What is the sum of the areas of these triangles?

Problem 2

Find the product of the roots of $\sqrt{30}x^{\log_{30}x} = x^2$.

Problem 3

Let $ABCD$ be a square of side length 2. Let E be the midpoint of $\overline{AB}$, F be the midpoint of $\overline{BC}$, and $\overline{AF}$ intersects $\overline{DB}$ at G and intersects $\overline{DE}$ at H. Then the area of quadrilateral $BEHG$ is $P\%$ of the area of square $ABCD$. What is P rounded to the nearest tenth?

Problem 4

Find the remainder when

$$1^{2018} + 2^{2018} + 3^{2018} + \cdots + 2018^{2018}$$

is divided by 7.

Problem 5

A regular hexagon is divided into six regions of equilateral triangles of the same size. Brandon plans to paint each equilateral
triangle with one of the colors chosen from red, green, blue,
yellow, and black, and doesn't want to paint any two adjacent
triangles with the same color. In how many ways can Brandon
paint the whole hexagon?

Problem 6

For real number r, let $\lfloor r \rfloor$ represent the greatest integer not exceeding r, and define $\{r\} = r - \lfloor r \rfloor$ (in other words, $\{r\}$ is the
fractional part of r). Let x be a real number and $1 \leq x < 7$, how
many solutions does the following equation have?

$$\{x^2\} = \{x\}^2.$$

Problem 7

In trapezoid $ABCD$, $\angle A = \angle D = 90°$. The sides have length
$AB = 9$, $BC = CD = 17$, and $AD = 15$. A circle through points
B and D intersects the extension of $\overrightarrow{BA}$ at E, and intersects the
extension of $\overrightarrow{CB}$ at F. Find the value of $BE - BF$. Round your
answer to the nearest integer if necessary.

Problem 8

Let x, y, and z be positive real numbers satisfying

$$\begin{cases} x^2 + xy + y^2 = 25, \\ y^2 + yz + z^2 = 49, \\ z^2 + zx + x^2 = 64. \end{cases}$$

The value of x can be expressed as $\dfrac{m}{\sqrt{n}}$, where m and n are positive integers and n is not a multiple of the square of any primes. Find $m + n$.

Problem 9

From the list of positive integers $1, 2, 3, \ldots$, remove the numbers that are multiples of 3 or 4 but not multiples of 5. In the remaining list of numbers, what is the 2000th number?

Problem 10

Given $\triangle ABC$ with area 2, from the three sides construct squares $ABDE$, $CAFG$, and $BCHK$ on the outside of $\triangle ABC$. The lengths EF, GH, and KD can be used to form a triangle. Find the area of this triangle, rounded to the nearest integer if necessary.

Problem 11

Real number x satisfy the following equation:

$$\frac{x-7}{\sqrt{x-3}+2} + \frac{x-5}{\sqrt{x-4}+1} = \sqrt{10}.$$

Find the sum of all possible values of x.

Problem 12

In a cube, the 8 vertices, centers of the 6 faces, midpoints of the 12 edges, and the center of the cube, are a collection of 27 points. Among these 27 points, how many collinear triples are there? (A collinear triple consists of 3 points that are on the same line.)

Problem 13

Let x, y be real numbers satisfying

$$(x + \sqrt{x^2 - 2018})(y + \sqrt{y^2 - 2018}) = 2018.$$

Find the value of $2019xy - 2018y^2 + 2017x^2 - 2017^2$.

Problem 14

In $\triangle ABC$, $\angle ABC = \angle ACB = 40°$. Let P be an interior point in $\triangle ABC$, such that $\angle PAC = 20°$, $\angle PCB = 30°$.

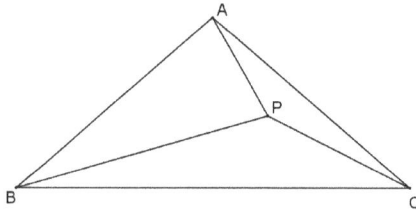

Find the degree measure of $\angle BPC$.

Problem 15

Find the smallest M such that the inequality

$$2ab + 3bc + 2cd \leq M(a^2 + b^2 + c^2 + d^2)$$

holds for all groups of real numbers a, b, c, d.

Problem 16

Start with right triangular prism $ABC - DEF$ with volume 1. Divide the prism into four solids (which combine to form the full prism) by cutting the prism along the plane through points A, B, F and the plane through points D, E, C. The volume of the largest of the solids can be written as $\dfrac{P}{Q}$ for positive integers P, Q with $\gcd(P, Q) = 1$. What is $P + Q$?

Problem 17

Let x and y be integers satisfying

$$\frac{x+y}{x^2 - xy + y^2} = \frac{3}{7}.$$

Find the sum of all possible values of y.

Problem 18

The unit cube is the collection of points (x, y, z) where $0 \leq x \leq 1$, $0 \leq y \leq 1$, and $0 \leq z \leq 1$. The three planes $x = y$, $y = z$, and $z = x$ divide the unit cube into several non-overlapping polyhedra. How many such polyhedra are there?

Problem 19

There are 2018 points in the interior of $\triangle ABC$, where no three of the points (including A, B, and C) are collinear. Using these points plus the points A, B, C as vertices, how many non-overlapping triangles can be formed?

Problem 20

What is the remainder when $47^{43} + 49^{46}$ is divided by 2303?

1.6 ZIML March 2018 Varsity Division

Below are the 20 Problems from the Varsity Division ZIML Competition held in March 2018.
The answer key is available on p.208 in the Appendix.
Full solutions to these questions are available starting on p.131.

Problem 1
Arrange the letters "$SEEMTEEM$" in a row where no consecutive letters can be the same. How many arrangements are possible?

Problem 2
Given that $i = \sqrt{-1}$, and

$$i + 2i^2 + 3i^3 + \cdots + 99i^{99} + 100i^{100} = a + bi,$$

where a and b are real numbers. Find $|a| + |b|$.

Problem 3
The sum of n $(n > 1)$ consecutive positive integers is 6^3. Find the sum of all possible n.

Problem 4
Find the constant term of

$$\left(1 + x + \frac{1}{x^2}\right)^{10}.$$

Problem 5

In $\triangle ABC$, $AB = AC$. The median $\overline{BE}$ divides the perimeter of $\triangle ABC$ into two parts of lengths 12 and 15. Find the sum of all possible lengths of $\overline{AB}$.

Problem 6

Let $\phi(n)$ denote Euler's totient function, which means for any positive integer n, $\phi(n)$ equals the number of positive integers m such that $1 \le m \le n$ and $\gcd(n,m) = 1$. How many positive integers n less than 2000 satisfy that $\phi(n) = \dfrac{n}{2}$?

Problem 7

Let α and β be acute angles and $\alpha + \beta = 90°$, also $\sin\alpha$ and $\sin\beta$ are the roots of the equation $2x^2 - 2\sqrt{2}x + c = 0$. Suppose α equals k degrees, What is the value of $c + k$?

Problem 8

The sum of 100 positive integers is 101101. What is the largest possible greatest common divisor of these 100 numbers?

Problem 9

A rectangular path of size 7×1 is to be paved with identical 1×1 white tiles and identical 2×1 black tiles. In how many ways can the path be paved?

Problem 10

Given that α and β are acute angles, satisfying

$$\cos\alpha + \cos\beta - \cos(\alpha+\beta) = \frac{3}{2},$$

find the sum of all possible values of $\alpha + \beta$ in degrees.

Problem 11

Let D-ABC be a regular tetrahedron. Let E, F be the midpoints of $\overline{DC}$ and $\overline{AB}$ respectively. Find the degree measure of the angle between skew lines $\overline{EF}$ and $\overline{DA}$.

Problem 12

Let a, b, c be the three sides of scalene $\triangle ABC$, satisfying (1) b is an integer; (2) a, b, c is an arithmetic progression; (3) $a^2 + b^2 + c^2 = 84$. Find the value of b.

Problem 13

Let n be a positive integer and

$$\sin\frac{\pi}{2n+1}\sin\frac{2\pi}{2n+1}\cdots\sin\frac{n\pi}{2n+1} = \frac{3}{16},$$

Find n.

Problem 14

In $\triangle ABC$, $AB = AC$. Let D be a point on $\overline{AC}$ so that $\triangle ABD$ and $\triangle BCD$ are both isosceles triangles. Find the sum of all possible measures of $\angle BAC$ in degrees, rounded to the nearest tenth if necessary.

Problem 15

Let
$$(1-x+x^2)^{19} = 1 + ax + bx^2 + \cdots + x^{38},$$
find $a+b$.

Problem 16

Find the largest integer not exceeding the following sum:

$$1 + \frac{1}{\sqrt{2}} + \frac{1}{\sqrt{3}} + \frac{1}{\sqrt{4}} + \cdots + \frac{1}{\sqrt{168}}.$$

Problem 17

You throw a dart towards a target 9 times, hitting 4 times. The probability that exactly 2 of those hits are consecutive is $\frac{m}{n}$, where m and n are relatively prime positive integers. Find $m+n$.

Problem 18

On a sheet of metal in the shape of a circle with radius 1 (shown in the left of the diagram), cut out a sector with angle α, and use the remaining piece to form a cone (shown in the right of the diagram). When $\alpha = a - b\sqrt{c}$ degrees, the cone has the maximum volume. Here a, b, c are positive integers and c does not have square factors greater than 1. Calculate $a + b + c$.

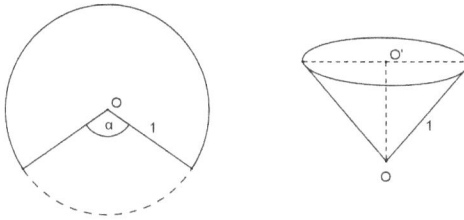

Problem 19

Find the last 6 digits of the following number:

$$1999^{1999^{1999^{\cdot^{\cdot^{1999}}}}}$$

where 1999 appears 1999 times in total.

Problem 20

Let x, y, z, t be real numbers satisfying

$$\frac{x}{y+z+t} = \frac{y}{z+t+x} = \frac{z}{t+x+y} = \frac{t}{x+y+z},$$

find the sum of all possible values of:

$$\left| \frac{x+y}{z+t} + \frac{y+z}{t+x} + \frac{z+t}{x+y} + \frac{t+x}{y+z} \right|.$$

1.7   ZIML April 2018 Varsity Division

Below are the 20 Problems from the Varsity Division ZIML
Competition held in April 2018.
The answer key is available on p.209 in the Appendix.
Full solutions to these questions are available starting on p.145.

Problem 1
Consider $\triangle ABC$, and D, E, F points on AC such that $AE : ED =$
$1 : 1$ and $DF : FC = 1 : 1$. Extend CB and AB to points G and H
such that EG and FH are parallel to DB. $\dfrac{[ABC]}{[DHG]} = \dfrac{P}{Q}$ in simplest
terms. What is $P + Q$?

Problem 2
Let
$$N = \binom{2017}{0} - \binom{2017}{2} + \binom{2017}{4} - \cdots + \binom{2017}{2016},$$
find the value of $\log_2 N$.

Problem 3
Find the remainder when $(7^{2019} + 46)^{2018}$ is divided by 50.

Problem 4

Let a, b, c be rational numbers, not all 0, such that the three roots of the equation

$$x^3 + ax^2 + bx + c = 0$$

are in fact a, b, and c. Find the sum of all possible values of $a^2 + b^2 + c^2$, rounded to the nearest integer if necessary.

Problem 5

Consider fractions $\dfrac{N}{M}$ for positive M, N with $\gcd(M, N) = 1$ with decimal form $0.\overline{abcd}$ and distinct a, b, c, d. For example, $0.\overline{1234} = 0.123412341234\ldots$ is such a number but $0.\overline{1212} = 0.121212\ldots$ is not.

What is the minimum value of $N + M$ over all such fractions?

Problem 6

In $\triangle ABC$, D is a point on $\overline{AC}$, and $AB = \sqrt{6}$, $AD = 2$, and $DC = 1$. Also given that $\angle ADB = 60°$. Find the measure of $\angle CBD$ in degrees.

Problem 7

Suppose you toss an unfair (two-sided) coin where the probability of getting heads on a given flip is $K\%$. You then roll a fair die once if you get heads and twice if you get tails. If the probability the coin came up heads given the sum of your rolls is 3 is 50%, what is K, rounded to the nearest integer if necessary.

Problem 8

Let $\dfrac{m}{n}$ be the smallest possible ratio between a 3-digit number (in base 10) and the sum of its digits, where m and n are positive integers and $\gcd(m,n) = 1$. What is $m+n$?

Problem 9

How many numbers ≤ 1000 can be written in the form

$$r \cdot 40 + s \cdot 64$$

for non-negative integers r, s?

Problem 10

Let $ABCD\text{-}A_1B_1C_1D_1$ be a rectangular prism, and $AB = 5$, $BC = 4$, $BB_1 = 3$. Find the distance between skew lines $\overline{AB}$ and $\overline{DB_1}$, rounded to the nearest tenth if necessary. (Note: the distance between two skew lines is the shortest distance between any two points, one on each line.)

Problem 11

Let $f(x)$ be a function defined on the real numbers. Given that $f(0) = 5$, and

$$f(x+1)(1 - f(x)) = 1 + f(x)$$

for all $x \in \mathbb{R}$. Find the value of $f(2018)$, rounded to the nearest tenth if necessary.

Problem 12

Let $\dfrac{m}{n}$ be a rational number such that $\mathrm{lcm}(m,n) = 2018$. How many possible such fractions are (strictly) between 0 and 1?

Problem 13

Let p, q, r be real numbers, satisfying

$$\arctan p + \arctan q + \arctan r + \frac{\pi}{2} = 0,$$

find the sum of all possible values of $pq + qr + rp$, rounded to the nearest tenth if necessary.

Problem 14

Let D and E be points on segment $\overline{AB}$, where D is between A and E, and $AD = DE = EB$. Construct a circle using $\overline{DE}$ as diameter, and let C be a point on the circle. Let $\alpha = \angle ACD$ and $\beta = \angle BCE$, find the value of $\tan\alpha \cdot \tan\beta$, rounded to the nearest hundredth if necessary.

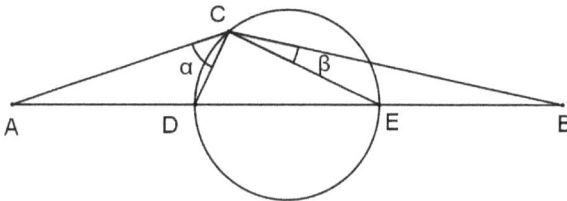

Problem 15

Find all possible sets of three prime numbers whose product is eleven times their sum. The ratio of the product for the largest set to the product for the smallest set can be written as $\dfrac{M}{N}$ for positive integers M, N with $\gcd(M,N) = 1$. What is $M + N$? (If there are no such sets, input an answer of 0; if there is one such set, input an answer of 1.)

Problem 16

Find the constant term of

$$\left(\sqrt{x} + \frac{1}{4\sqrt{x}} - 1 \right)^6.$$

Round your answer to the nearest integer.

Problem 17

Find the maximum value of

$$\sqrt{12x - x^2 - 11} + \sqrt{68x - x^2 - 256}$$

and round the answer to the nearest tenth if necessary.

Problem 18

Find the remainder when 61! is divided by 71.

Problem 19

In a convex polygon with n sides, any three diagonals do NOT go through the same point. The diagonals are divided into a total of 704 pieces by the intersections of all the diagonals. What is n?

Problem 20

In $\triangle ABC$, $AB = AC = 5$. On side $\overline{BC}$ there are 40 points

$$P_1, P_2, \ldots, P_{40},$$

in that order, such that

$$BP_1 = P_1 P_2 = P_2 P_3 = \cdots = P_{39} P_{40} = P_{40} C.$$

Define
$$k_i = AP_i^2 + BP_i \cdot CP_i \text{ for } i = 1, 2, \ldots, 40.$$

Find the value of $k_1 + k_2 + \cdots + k_{40}$.

1.8 ZIML May 2018 Varsity Division

Below are the 20 Problems from the Varsity Division ZIML
Competition held in May 2018.
The answer key is available on p.210 in the Appendix.
Full solutions to these questions are available starting on p.158.

Problem 1
In the sequence of increasing positive integers

$$2, 3, 5, 6, 7, 10, 11, 12, 13, 14, 15, 17, \ldots$$

all positive integers are included except for perfect squares and
perfect cubes. Find the 800th number in this sequence.

Problem 2
Let a and b be digits, satisfying $\overline{25ab} = 2^5 \cdot a^b$ ($\overline{25ab}$ is a 4-digit
number). Find $10a + b$.

Problem 3
Let $f(x) = 2x^2 - 2x + k$, and

$$\log_2 f(a) = 3, \quad f(\log_2 a) = k, \quad a > 0, \quad a \neq 1,$$

find the minimum value for $f(\log_2 x)$, express your answer in
decimal, rounded to the nearest tenth if necessary.

Problem 4

Let $ABCD$-$A'B'C'D'$ be a cube. A plane cut through this cube. Suppose the angles formed by this plane with the 12 edges of cube $ABCD$-$A'B'C'D'$ are all equal. Let this angle be θ, find the value of $\csc^2\theta$.

Problem 5

What is the sum of the distinct prime factors of the 12-digit number 999999999999?

Problem 6

Let $k \in [-1,1]$, suppose the straight line $y = \dfrac{\pi}{4}$ and the curve represented by the equation

$$(x - \arcsin k)(x + \arccos k) + (y - \arcsin k)(y - \arccos k) = 0$$

intersect at points M and N. Find the minimum possible length of $\overline{MN}$, rounded to the nearest hundredth if necessary.

Problem 7

Select 3 different numbers from $\{0,1,2,3,4,5,6,7,8,9\}$, so that their sum is an even number greater or equal to 10. How possible choices are there?

Problem 8

The region R in the complex plane contains all points z such that both $z/20$ and $20/\bar{z}$ have real and imaginary parts between 0 and 1, inclusive. Find the area of R, using 3.14 as the value of π, and round your answer to the nearest integer if necessary.

Problem 9
Evaluate
$$\sin^2 80° + \sin^2 40° - \cos 50° \cos 10°,$$

express your answer in decimal, rounded to the nearest hundredth if necessary.

Problem 10
Solve the following equation:

$$|\log_{\sqrt{10}} x - 2| - |\log_{10} x - 2| = 2,$$

then find the difference between the largest solution and smallest solution, rounded to the nearest hundredth if necessary.

Problem 11
How many pairs of positive integers (a, b) satisfy the following: $1 \le a \le b \le 100$ and $7 \mid (a^2 + b^2)$?

Problem 12
A frog sits at the vertex A of a regular hexagon $ABCDEF$, and starts to jump. At each jump, the frog randomly gets to one of the two adjacent vertices. The frog stops if he reaches D in fewer than 5 jumps. If not, he stops after 5 jumps. From start to stop, how many possible paths can the frog jump?

Problem 13

Let a, b, c be nonnegative real numbers. Find the minimum value of

$$\frac{c}{a} + \frac{a}{b+c} + \frac{b}{c},$$

rounded to the nearest tenth if necessary.

Problem 14

Some positive integers can be written as the sum of two or more consecutive positive integers, such as $9 = 4 + 5$, and $26 = 5 + 6 + 7 + 8$. Find the number of positive integers less than 1000 that CANNOT be written as the sum of two or more consecutive positive integers.

Problem 15

Let a be a real number such that the ellipse

$$x^2 + 4\left(y - \frac{a}{2}\right)^2 = 4$$

and the parabola

$$x^2 = 2y$$

have at least one common point. Let M and N be the maximum and minimum possible values of a, respectively, find the difference $M - N$, rounded to the nearest hundredth if necessary.

Problem 16

Given 6 distinct colors, choose some of these colors to paint the faces of a cube. Each face is painted with one color, and faces that share a common edge are painted with different colors. Suppose two ways of painting are considered the same if they become identical after rotation, in how many different ways can the cube be painted?

Problem 17

Define sequence $a_n (n \geq 0)$ as follows:

$$a_n = \frac{(3+2\sqrt{2})^n + (3-2\sqrt{2})^n}{2}.$$

Find the last digit of a_{2018}.

Problem 18

Let positive integer n be a perfect square whose units digit is not 0. If the last two digits of n is crossed out, the remaining number is still a perfect square. What is the largest such number n?

Problem 19

In tetrahedron P-ABC, $\angle APB = \angle BPC = \angle CPA = 90°$, and the sum of the lengths of the six edges is 18. The maximum possible value for its volume is expressed as $K\sqrt{M} - N$, where K, M, and N are positive integers, and M contains no square factors greater than 1. Find $K + M + N$.

Problem 20

Suppose you need to print 3-digit numbers $100, 101, \ldots, 999$ on cards to display, one number per card. You also want to protect the environment and print as few cards as possible. You discover that some cards can be turned upside-down and still has a valid 3-digit number. For example, 168 can be turned into 891. and 806 can be turned into 908.. But some cards can't, such as 239. You want to reuse as many cards as possible by turning them upside-down. What is the minimum number of cards you should print so that all 3-digit number are still represented?

1.9   ZIML June 2018 Varsity Division

Below are the 20 Problems from the Varsity Division ZIML Competition held in June 2018.
The answer key is available on p.211 in the Appendix.
Full solutions to these questions are available starting on p.173.

Problem 1
Find the smallest real root of the following equation:

$$\sqrt{x^2 - 8x + 25} + \sqrt{x^2 - 10x + 29} = \sqrt{26}.$$

Express the answer as a decimal, rounded to the nearest tenth if necessary.

Problem 2
Let $ABCD\text{-}A_1B_1C_1D_1$ be a cube and $AB = \sqrt{3}$. Let M be the center of the face $A_1B_1C_1D_1$, and N be the center of the face BCC_1B_1. Find the distance between the skew lines $\overline{AN}$ and $\overline{DM}$.

Problem 3

Assume 40% of the population suffers from seasonal allergies. At their medical lab, Jason is helping develop a screening test for allergies. Their test currently has a 5% false positive rate, meaning that 5% of people who do not suffer from allergies test positive with their test. Jason wants their test to be designed so that $\geq 90\%$ of people who test positive actually suffer from allergies. For this goal, their test must have a false negative rate of $\leq K\%$. (Recall a false negative is when someone suffers from allergies but tests negative.) What is K, rounded to the nearest tenth if necessary?

Problem 4

Let $\overline{abcd}$ be a four-digit positive integer, and the square of the sum of its first two digits and its last two digits is equal to the four-digit number itself, that is,

$$(\overline{ab} + \overline{cd})^2 = \overline{abcd}.$$

Find the sum of all such four-digit numbers.

Problem 5

In $\triangle ABC$, $\overline{BD}$ is the altitude on side $\overline{AC}$. Let $BD = h$, $BC = a$, $CA = b$, and $AB = c$, and the lengths h, a, b, c form an arithmetic progression with common difference d. Find the value of $\dfrac{h}{d}$, rounded to the nearest integer if necessary.

Problem 6

Let $a_1, a_2, \ldots, a_{101}$ be a sequence of 101 consecutive positive integers, satisfying

$$a_1^2 + a_2^2 + \cdots + a_{51}^2 = a_{52}^2 + a_{53}^2 + \cdots + a_{101}^2.$$

Find the minimum possible value of a_1.

Problem 7

Call a sequence $(a_1, a_2, \ldots, a_n)$ pyramidal if there is a k such that $1 \leq k \leq n$ and $a_1 \leq \cdots \leq a_k \geq \cdots \geq a_n$. For example, $(1, 2, 3, 4)$, $(1, 2, 3, 1)$, $(2, 1)$, and (1) are all pyramidal but $(2, 1, 2)$ is not. How many pyramidal sequences made up of positive integers are there such that the sum of all terms in the sequence is 7?

Problem 8

Let b be a positive integer less than 100. In base b, the number 2101 is a perfect square. Find the sum of all possible values of b.

Problem 9

The regular octagon shown below has 8 lines of symmetry (reflections).

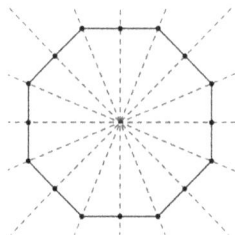

Color the 8 vertices and 8 midpoints of the edges of this octagon using 4 colors (it is not necessary to use all the colors). How many such colorings are there so that only the horizontal and vertical symmetries are preserved?

Problem 10

Suppose a is an integer, and the equation

$$ax^2 - (a-3)x + a - 2 = 0$$

has at least one integer root for x. Find the sum of all possible values of a.

Problem 11

Let a, b, c be three positive integers, all greater than 1, satisfying

$$a \mid (bc+1), \quad b \mid (ca+1), \quad c \mid (ab+1).$$

Find the largest possible value of the product abc.

Problem 12

In rectangle $ABCD$, $AB = 4$, $AD = 8$. Fold the along diagonal $\overline{BD}$ so that point C becomes point C'. Connect $\overline{BC'}$, intersecting $\overline{AD}$ at E. Find the area of $\triangle BED$, rounded to the nearest integer if necessary.

Problem 13

Find the sum of all real roots of the following equation:

$$\sqrt[3]{x^2 - 4} + 4 = 2(\sqrt[3]{x+2} + \sqrt[3]{x-2}).$$

Problem 14

Let $ABCD$ be a rhombus, $AC > BD$, satisfying

$$\frac{AC}{AB} = \frac{AB}{BD}.$$

Find the measure of $\angle DAB$ in degrees.

Problem 15

There are two distinct arrangements of 4 queens on a 4×4 board so that no two queens are in the same row, column, or diagonal:

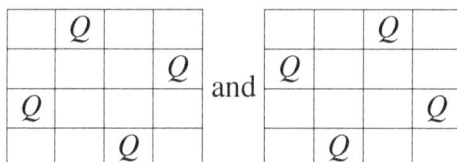

$$\begin{array}{|c|c|c|c|}\hline & Q & & \\\hline & & & Q \\\hline Q & & & \\\hline & & Q & \\\hline\end{array} \text{ and } \begin{array}{|c|c|c|c|}\hline & & Q & \\\hline Q & & & \\\hline & & & Q \\\hline & Q & & \\\hline\end{array}$$

How many distinct arrangements of 3 queens on a 4×4 board are there?

Problem 16

Sequence $\{a_k\}$ is defined as follows: $a_1 = 2$, $a_2 = 0$, $a_3 = 1$, $a_4 = 8$, and

$$a_{i+4} = a_i + a_{i+3}, \quad i \geq 1.$$

Find the remainder when $a_{2018}^2 + a_{2019}^2 + a_{2020}^2$ is divided by 4.

Problem 17

You have 7 identical red cards, 7 identical white cards, and 7 identical blue cards. The cards are arranged in a row such that no two red cards are adjacent. Both the white cards and blue cards are arranged in 3 separate groups. (For reference, we would say $WWBWBBBWWWW$ has 3 groups of white and 2 groups of blue.) How many such arrangements are there?

Problem 18

Let $f(\theta) = (1 + \cos \theta)^2 \sin^2 \theta$. Find the maximum value of $f(\theta)$, rounded to the nearest hundredth if necessary.

Problem 19

Let x and y be integers satisfying

$$x^2 + xy + 4y^2 = 61.$$

Find the largest possible value of $|x| + |y|$.

Problem 20

Let x and y be positive real numbers, and define

$$z = \min\left\{x, \frac{y}{x^2 + y^2}\right\}.$$

Find the maximum possible value of z^2. Express the answer in decimal, rounded to the nearest tenth if necessary.

2. ZIML Solutions

This part of the book contains the official solutions to the problems from the nine Varsity Division ZIML Contests from the 2017-18 School Year.

Students are encouraged to discuss and share their own methods to the problems using the Discussion Forum on ziml.areteem.org.

2.1 ZIML October 2017 Varsity Division

Below are the solutions from the Varsity Division ZIML Competition held in October 2017.
The problems from the contest are available on p.17.

Problem 1 Solution
First note ℓ_1 and ℓ_2 are parallel, with ℓ_3 perpendicular. As ℓ_3
intersects ℓ_1 and ℓ_2 are $(4,2)$ and $(2,4)$ we have that the distance
between ℓ_1 and ℓ_2 is

$$\sqrt{(4-2)^2 + (2-4)^2} = 2\sqrt{2}.$$

Hence any circle tangent to ℓ_1 and ℓ_2 must have radius $\sqrt{2}$
with center on the line $y = x$ (the perpendicular bisector of
$(4,2),(2,4)$). $(x-3)^2 + (y-3)^2 = 2$ is the circle of all points distance $\sqrt{2}$ from $(3,3)$. This circle intersects the line $y = x$ at $(2,2)$
and $(4,4)$, so the two circles that are tangent to all three lines are
$(x-2)^2 + (y-2)^2 = 2$ and $(x-4)^2 + (y-4)^2 = 2$. Hence the
smallest value of $H + K + S = 2 + 2 + 2 = 6$.

Answer: 6

Problem 2 Solution

$$
\begin{aligned}
&\quad\ (57^{37} + 46)^{26} \\
&\equiv\ (7^{37} - 4)^{26} = (7 \times (7^2)^{18} - 4)^{26} \\
&\equiv\ (7 \cdot (-1)^{18} - 4)^{26} = (7 - 4)^{26} = 3^{26} = 3 \cdot (3^5)^5 \\
&\equiv\ 3 \cdot (-7)^5 = -3 \cdot 7 \cdot (7^2)^2 \\
&\equiv\ -21 \\
&\equiv\ 29 \quad (\mathrm{mod}\ 50).
\end{aligned}
$$

Answer: 29

Problem 3 Solution
Case (1): the first digit is 3. There are $4! = 24$ numbers.

Case (2): the first digit is not 3, so the first digit has three choices, and the hundreds digit has three choices, and then the remaining 3 digits have $3! = 6$ choices, so that is $3 \cdot 3 \cdot 6 = 54$ choices.

Combining the two cases, $24 + 54 = 78$ numbers can be formed.

Answer: 78

Problem 4 Solution
By Vieta's formulas,

$$x_1^2 + x_2^2 = (x_1 + x_2)^2 - 2x_1x_2 = a^2/4 - b = 1,$$

so $a^2 = 4 + 4b$. The discriminant then satisfies

$$\Delta = a^2 - 8b = 4 - 4b \geq 0,$$

so $b \leq 1$. We also have $4 + 4b = a^2 \geq 0$, so $-1 \leq b$. Therefore $-1 \leq b \leq 1$ (and for each such b, $a = \pm\sqrt{4 + 4b}$) so

$$L - K = 1 - (-1) = 2.$$

Answer: 2

Problem 5 Solution
Using the identity

$$\frac{1}{k}\binom{n-1}{k-1} = \frac{1}{n}\binom{n}{k},$$

and let $n = 10$, we get

$$\frac{1}{k}\binom{9}{k-1} = \frac{1}{10}\binom{10}{k}.$$

Let $k = 1, 2, 3, \ldots, 10$ and add up the terms,

$$\binom{9}{0} + \frac{1}{2}\binom{9}{1} + \frac{1}{3}\binom{9}{2} + \frac{1}{4}\binom{9}{3} + \cdots + \frac{1}{10}\binom{9}{9}$$
$$= \frac{1}{10} \sum_{k=1}^{10} \binom{10}{k}$$
$$= \frac{2^{10} - 1}{10}$$
$$= \frac{1023}{10}.$$

Therefore $a = 1023, b = 10$, and $a + b = 1033$.

Answer: 1033

Problem 6 Solution

Let A, B, C, D, E be the events that box $1, 2, 3, 4, 5$ (respectively) get at least 3 balls. We want $n(A \cup B \cup C \cup D \cup E)$. Note

$$n(A) = \cdots = n(E) = \binom{5 + 5 - 1}{5}$$

using stars and bars (put 3 balls in the respective box, and then arrange the remaining $8 - 3 = 5$ balls in any of the boxes). Similarly we have

$$n(A \cap B) = \cdots = n(D \cap E) = \binom{2 + 5 - 1}{2}.$$

Since the intersection of 3 or more of these sets is empty,

$$n(A \cup B \cup C \cup D \cup E) = \binom{5}{1} \cdot \binom{9}{5} - \binom{5}{2} \cdot \binom{6}{2} = 480,$$

as our final answer.

Answer: 480

Problem 7 Solution

Extend $\overline{AD}$ and $\overline{BC}$ to intersect at E. Then the two triangles $\triangle ABE$ and $\triangle CDE$ are both 30-60-90 triangles, so

$$AE = 2AB = 8, DE = 3.$$

Thus

$$CD = \sqrt{3}, CE = 2\sqrt{3},$$

also

$$BE = 4\sqrt{3}, BC = 2\sqrt{3},$$

therefore $BC/CD = 2$.

Answer: 2

Problem 8 Solution

Let $u = \sqrt[3]{5-x}$ and $v = \sqrt{x-4}$, then $u+v = 1$ and

$$u^3 + v^2 = 5 - x + x - 4 = 1.$$

From $u+v = 1$, we get $v = 1 - u$, so

$$u^3 + u^2 - 2u + 1 = 1,$$

thus

$$u^3 + u^2 - 2u = 0,$$

so

$$u(u-1)(u+2) = 0,$$

then $u = 0$ or $u = 1$ or $u = -2$. Therefore $x = 5$ or $x = 4$ or $x = 13$, and they are all roots after checking. Thus the sum of these roots is 22.

Answer: 22

Problem 9 Solution

Let x_1 and x_2 be the two positive integer roots, then $x_1 + x_2 = -a$

and $x_1 x_2 = 1 - b$, so a and b must also be integers, and

$$
\begin{aligned}
a^2 + b^2 &= (x_1 + x_2)^2 + (1 - x_1 x_2)^2 \\
&= (1 + x_1^2)(1 + x_2^2) \\
&= 629 \\
&= 17 \cdot 37.
\end{aligned}
$$

Since $x_1, x_2 \geq 1$, we have (without loss of generality) $1 + x_1^2 = 17$ and $1 + x_2^2 = 37$. Since x_1 and x_2 are positive, $x_1 = 4$ and $x_2 = 6$. Hence $a = -(x_1 + x_2) = -10$ and $b = 1 - x_1 x_2 = -23$. Therefore $|a| + |b| = 10 + 23 = 33$.

Answer: 33

Problem 10 Solution

It can be proven that ray DF is the bisector of $\angle ADB$. Let G be the intersection point of the bisectors of triangle DAB.

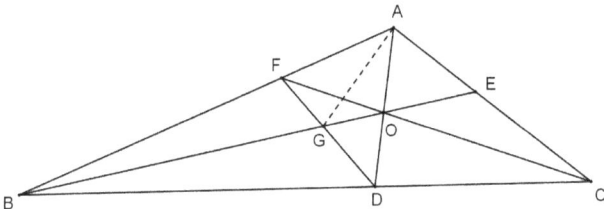

Since

$$
\angle DGB + \frac{1}{2} \angle ABD + \frac{1}{2} \angle ADB = 180°,
$$

we have

$$
\angle DGB + \frac{1}{2}(180° - \angle BAD) = 180°,
$$

then

$$
\angle DGB = 90° + \frac{1}{2} \angle BAD = 90° + 30° = 120°.
$$

Thus $\angle FGO = \angle DGB == 120°$. Hence,

$$\angle FGO + \angle FAO = 180°,$$

which means quadrilateral $AOGF$ is cyclic. Therefore,

$$\angle DFO = \angle GFO = \angle GAO = 30°.$$

Answer: 30

Problem 11 Solution

We use complementary counting. The total numbers of ways without restriction is $\binom{9}{3} = 84$. So we want to choose 3 numbers a, b, c, in increasing order, with no consecutive numbers. Let

$$a' = a, \quad b' = b - 1, \quad c' = c - 2.$$

So

$$1 \leq a' < b' < c' \leq 7.$$

The numbers a', b', c' do not need to be separated, and the number of ways to choose a', b', c' out of 7 is the same as the number of ways to choose a, b, c out of 9. Thus there are $\binom{7}{3} = 35$ ways. Thus there are $84 - 35 = 49$ ways to choose 3 numbers with at least one pair of consecutive numbers.

Answer: 49

Problem 12 Solution

Let $a = 24m$ and $b = 24n$ where $\gcd(m, n) = 1$, and then

$$24mn = 144, \quad 24m + 24n = 5.$$

Solve for m, n knowing that $mr < n$, we get $m = 2, n = 3$, so a, b are 48 and 72, and $100a + b = 4872$.

Answer: 4872

Problem 13 Solution

There are $\binom{10}{4} = 210$ groups of 4 points. We shall take away the groups whose all 4 points are on the same plane. There are 3 cases of such groups:

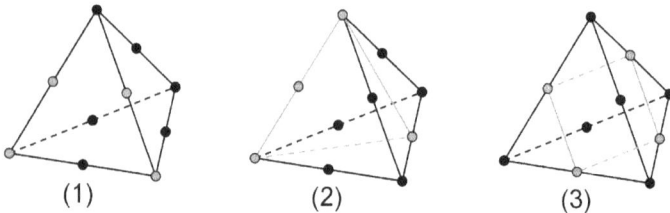

(1)   (2)   (3)

Case (1) All 4 points are on the same face of the tetrahedron: $4 \cdot \binom{6}{4} = 60$ groups;

Case (2) 3 points on the same edge plus 1 midpoint on the opposite edge: 6 groups;

Case (3) Midpoints of edges that form parallelograms: 3 groups. Taking these away, the remaining groups consists of non-coplanar points. So the final answer is

$$210 - 60 - 6 - 3 = 141.$$

Answer: 141

Problem 14 Solution

We get that $M = \dfrac{2017 \times 2016}{2} = 2017 \times 1008$, and since 2017 is prime, $\gcd(2017, 1008) = 1$.

By Wilson's Theorem,

$$2016! \equiv -1 \equiv 2016 \pmod{2017}.$$

Also we know that

$$2016! \equiv 0 \equiv 2016 \pmod{1008},$$

therefore by Chinese Remainder Theorem,

$$2016! \equiv 2016 \pmod{2017 \times 1008}.$$

Answer: 2016

Problem 15 Solution

This is the same as $z^3 + z^4 + z^6 + z + z^2 + z^5$. We have

$$z^6 + z^5 + z^4 + z^3 + z^2 + z + 1 = 0,$$

so the answer is -1.

Answer: -1

Problem 16 Solution

Let $k = ab - cd$, so $a = xk, b = yk, c = zk, d = wk$, and then

$$k = ab - cd = xyk^2 - zwk^2 = (xy - zw)k^2,$$

which means $k^2 \mid k$. Since k is a positive integer, the only way this can happen is that $k = 1$.

Answer: 1

Problem 17 Solution

First note $\dfrac{BO}{OD} = \dfrac{[BOA]}{[DOA]} = \dfrac{[BOC]}{[DOC]}$ as the triangles share the same height. Hence adding these ratios we get

$$\frac{BO}{OD} = \frac{[BOA] + [BOC]}{[DOA] + [DOC]} = \frac{[ABC]}{[ACD]} = \frac{1}{2},$$

so $\dfrac{BO}{BD} = \dfrac{BO}{BO+OD} = \dfrac{1}{3}$. Also, $\dfrac{[ABO]}{[ABD]} = \dfrac{BO}{BD} = \dfrac{1}{3}$, thus

$$[ABO] = 6 \cdot \dfrac{1}{3} = 2.$$

Answer: 2

Problem 18 Solution

We rotate $\triangle DAP$ about point D by $90°$ so that the side $\overline{AD}$ coincides with side $\overline{CD}$, and point P reaches point P', as shown. We know that $\triangle DP'P$ is an isosceles right triangle, and $\angle DP'P = 45°$. Let $\angle PP'C = \theta$.

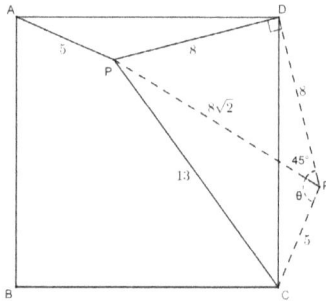

Applying Law of Cosines on $\triangle PCP'$,

$$13^2 = 5^2 + (8\sqrt{2})^2 - 2 \times 5 \times 8\sqrt{2}\cos\theta,$$

thus

$$\cos\theta = \dfrac{25 + 128 - 169}{2 \times 5 \times 8\sqrt{2}} = -\dfrac{\sqrt{2}}{10}.$$

Hence

$$\sin\theta = \sqrt{1 - \cos^2\theta} = \dfrac{7\sqrt{2}}{10}.$$

To calculate the area of $ABCD$, we use Law of Cosines again,

$$
\begin{aligned}
[ABCD] = CD^2 &= 8^2 + 5^2 - 2 \times 8 \times 5 \cos \angle DP'C \\
&= 89 - 80 \cos(45° + \theta) \\
&= 89 - 80(\cos 45° \cos \theta - \sin 45° \sin \theta) \\
&= 89 - 80 \left(\frac{\sqrt{2}}{2} \cdot \left(-\frac{\sqrt{2}}{10} \right) - \frac{\sqrt{2}}{2} \cdot \frac{7\sqrt{2}}{10} \right) \\
&= 89 + 8 + 56 \\
&= 153.
\end{aligned}
$$

Answer: 153

Problem 19 Solution

We know that

$$
\begin{aligned}
\log_3 x_1 + x_1 &= 3 \\
3^{x_2} + x_2 &= 3.
\end{aligned}
$$

The second equation above can be written as

$$
\log_3(3^{x_2}) + 3^{x_2} = 3.
$$

For $x > 0$, $f(x) = \log_3 x + x$ is an increasing function, and since $f(x_1) = f(3^{x_2}) = 3$, we have $x_1 = 3^{x_2}$. Therefore

$$
x_1 + x_2 = 3^{x_2} + x_2 = 3.
$$

Answer: 3

Problem 20 Solution

Pair up the odd numbers less than 10000 so that each pair has sum 10000. Let (a, b) be a pair, so $a + b = 10000$. Since

$$
(a+b) \mid (a^9 + b^9),
$$

we have

$$
10000 \mid a^9 + b^9.
$$

Since $f(a)$ is the number consisting of the last 4 digits of a^9, and $f(b)$ is the number consisting of the last 4 digits of b^9,

$$f(a) + f(b) = 10000.$$

Therefore, $f(a) > a$ if and only if $f(b) < b$, thus the sets A and B have the same number of elements. In other words, $|A| = |B|$.

Answer: 0

2.2 ZIML November 2017 Varsity Division

Below are the solutions from the Varsity Division ZIML Competition held in November 2017.
The problems from the contest are available on p.23.

Problem 1 Solution
All 9 rings are distinct, and since the order of the rings on each finger matters, this is eqiuvalent to just ordering all 9 rings. This can be done in $9! = 362880$ different ways.

Answer: 362880

Problem 2 Solution
Clearly $x = \pm 1$ are not solutions. If both sides of the equation are divided by $\sqrt[3]{x^2 - 1}$, we obtain

$$\sqrt[3]{\frac{x+1}{x-1}} + 2\sqrt[3]{\frac{x-1}{x+1}} = 3.$$

Let $y = \sqrt[3]{\dfrac{x+1}{x-1}}$, then

$$y + \frac{2}{y} = 3,$$

and so

$$y^2 + 2y - 3 = 0,$$

thus $y = 1$ or $y = -3$.

If $y = 1$, $\dfrac{x+1}{x-1} = 1$, and there is no solution for x. If $y = -3$, we get

$$\frac{x+1}{x-1} = -27,$$

so

$$x = \frac{13}{14}.$$

Therefore the answer is $13 + 14 = 27$.

Answer: 27

Problem 3 Solution
Since
$$2836 \equiv 4582 \equiv 5164 \equiv 6522 \pmod{m},$$
we obtain
$$1746 \equiv 582 \equiv 1358 \equiv 0 \pmod{m}.$$

This means m is a common factor of 1746, 582, and 1358. Since $\gcd(1746, 582, 1358) = 194$, $m \mid 194$. So m can be 2, 97, or 194. If $m = 2$, the remainder $r = 0$ which is not a solution; if $m = 97$, $r = 23$; if $m = 194$, $r = 120$. Hence the $r = 23$ or $r = 120$, so our answer is $23 + 120 = 143$.

Answer: 143

Problem 4 Solution
When we fold A to C we fold along the perpendicular bisector of $\overline{AC}$, giving the diagram below, where M is the midpoint of $\overline{AC}$ and $\triangle AME \cong \triangle AMF \cong \triangle CME \cong \triangle CMF$.

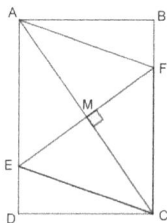

$CD = 5$ so letting $DE = BF = x$ we have $CE^2 = x^2 + 5^2$ but also $CE = BC - BF = 7 - x$. Thus
$$(7 - x)^2 = x^2 + 5^2 \Rightarrow 14x = 24 \Rightarrow x = \frac{12}{7} \Rightarrow 7 - x = \frac{37}{7}.$$

Therefore the area of the overlap region (the shaded region since we are calculating the overlap region after the folding) is

$$\frac{1}{2} \cdot \frac{37}{7} \cdot 5 = \frac{185}{14}.$$

This gives a final answer of $185 - 14 = 171$.

Answer: 171

Problem 5 Solution

Using the triangle inequality, we need (i) $2 + b > c$ or $c < b + 2$, (ii) $2 + c > b$ or $c > b - 2$, and (iii) $b + c > 2$ or $c > 2 - b$. Graphing this in a 3×4 rectangle, we get that (b, c) must be in the shaded region shown below:

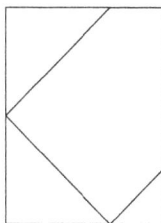

The full rectangle has area 12, while the shaded region has area $12 - 2 - 2 - \frac{1}{2} = 7.5$, so the probability is $\frac{7.5}{12} = 62.5\%$, so $L = 62.5$.

Answer: 62.5

Problem 6 Solution

Since $10000 = 2^4 \times 5^4$, its even positive divisors are of the form $2^a 5^b$ where $1 \le a \le 4$ and $0 \le b \le 4$, therefore the sum of these even divisors is

$$(2 + 2^2 + 2^3 + 2^4)(1 + 5 + 5^2 + 5^3 + 5^4) = 23430.$$

Answer: 23430

Problem 7 Solution

Note that $APBC$ is a cyclic quadrilateral by construction. Using Ptolemy's theorem, we know that

$$AC \cdot BP = AP \cdot BC + AB \cdot CP \Rightarrow BP = AP + CP$$

as $AB = BC = CA$ in the equilateral triangle. Hence

$$CP = BP - AP = 5 - 3 = 2.$$

Answer: 2

Problem 8 Solution

Since a, b, c, d are roots of

$$x^4 - 5x^2 + 3x + 1 = 0$$

or

$$x^4 = 5x^2 - 3x - 1$$

we can write

$$a^4 + b^4 + c^4 + d^4 = 5(a^2 + b^2 + c^2 + d^2) - 3(a + b + c + d) - 4.$$

We know $a + b + c + d = 0$ by Vieta's Formulas. Further we know

$$
\begin{aligned}
a^2 + b^2 + c^2 + d^2 \\
&= (a + b + c + d)^2 - 2(ab + ac + ad + bc + bd + cd) \\
&= 0 - 2(-5) \\
&= 10.
\end{aligned}
$$

Hence

$$
\begin{aligned}
a^4 + b^4 + c^4 + d^4 \\
&= 5(a^2 + b^2 + c^2 + d^2) - 3(a + b + c + d) - 4 \\
&= 5 \cdot 10 - 3 \cdot 0 - 4 \\
&= 46.
\end{aligned}
$$

Answer: 46

Problem 9 Solution

Use complementary counting. First, we can use stars and bars to find the total number of ways to distribute the tickets without restriction, which is $\binom{71+3-1}{3-1} = 2628$. Then consider the case we don't want: one group has more tickets than the other two combined. Select that one group (3 choices), and then give 36 tickets to that group first. Then distribute the remaining 35 tickets to the 3 groups as usual; using stars and bars again we see this can be done in $\binom{35+3-1}{3-1} = 666$. Thus, the tickets can be distributed in $2628 - 3 \cdot 666 = 630$ different ways.

Answer: 630

Problem 10 Solution

Say the medians are $\overline{AD}, \overline{BE}, \overline{CF}$, which divide $\triangle ABC$ into six equal area triangles. Extend $\overline{BE}$ 2 units further, giving a point H. Note $AGCH$ is a parallelogram, so $CH = 3$. Hence, $\triangle GCH$ is a right triangle with area 6. Note that $[GCH] = [ACG] = [ABC]/3$ and thus $[ABC] = 18$.

Answer: 18

Problem 11 Solution

The pattern of the last two digits of powers of 7: 07, 49, 43, 01. Thus we need to look for $7^{7^{.^{.^{7}}}}$ (mod 4) (now six 7s). Note this is equivalent to $(-1)^{\text{odd number}} \equiv 3$ (mod 4). Hence the last two digits are 43.

Answer: 43

Problem 12 Solution

The number of non-negative solutions to

$$a+b+c+d \leq 23$$

equals the number of solutions to

$$a+b+c+d+e = 23$$

where e is a non-negative integer. Using stars and bars this is

$$\binom{23+5-1}{23} = \frac{27!}{4!\,23!} = \frac{27 \cdot 26 \cdot 25 \cdot 24}{24} = 17550.$$

Answer: 17550

Problem 13 Solution

Consider the diagram:

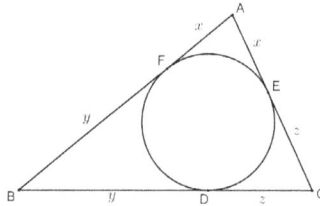

As shown, we set up a system of equations:

$$\begin{aligned} x+y &= 13; \\ y+z &= 14; \\ z+x &= 9. \end{aligned}$$

Adding the equations gives

$$2x+2y+2z = 36$$

so

$$x+y+z = 18.$$

Hence $z = 5, x = 4$ and $y = 9$, so

$$AF^2 + BD^2 + CE^2 = x^2 + y^2 + z^2 = 16 + 81 + 25 = 122.$$

Answer: 122

Problem 14 Solution

Simplify the argument first: let $x = \sqrt{3+\sqrt{5}} + \sqrt{3-\sqrt{5}}$, then

$$
\begin{aligned}
x^2 &= \left(\sqrt{3+\sqrt{5}} + \sqrt{3-\sqrt{5}}\right)^2 \\
&= 3+\sqrt{5} + 2\sqrt{3+\sqrt{5}} \cdot \sqrt{3-\sqrt{5}} + 3 - \sqrt{5} \\
&= 6 + 2\sqrt{9-5} \\
&= 10,
\end{aligned}
$$

so $x = \sqrt{10}$, and $\log_{10} x = \dfrac{1}{2} = 0.5$.

Answer: 0.5

Problem 15 Solution

Let O be the center of the regular n-gon, and assume

$$OA_1 = OA_2 = OA_3 = OA_4 = 1.$$

Then $\angle A_1 O A_2 = \dfrac{2\pi}{n}$, thus $A_1 A_2 = 2\sin\dfrac{\pi}{n}$.

Similarly, $A_1 A_3 = 2\sin\dfrac{2\pi}{n}$, and $A_1 A_4 = 2\sin\dfrac{3\pi}{n}$.

Since $\dfrac{1}{A_1 A_2} = \dfrac{1}{A_1 A_3} + \dfrac{1}{A_1 A_4}$, we get

$$\frac{1}{2\sin\dfrac{\pi}{n}} = \frac{1}{2\sin\dfrac{2\pi}{n}} + \frac{1}{2\sin\dfrac{3\pi}{n}}.$$

For simplicity, let $\theta = \dfrac{\pi}{n}$, and simplify,

$$\sin 2\theta \sin 3\theta = \sin \theta \sin 3\theta + \sin \theta \sin 2\theta.$$

hence (applying the sum-to-product formulas)

$$\begin{aligned}
\sin \theta \sin 3\theta &= \sin 2\theta \sin 3\theta - \sin \theta \sin 2\theta \\
&= \sin 2\theta (\sin 3\theta - \sin \theta) \\
&= 2 \sin 2\theta \cos 2\theta \sin \theta \\
&= \sin 4\theta \sin \theta.
\end{aligned}$$

Canceling $\sin \theta$, we get

$$\sin 3\theta = \sin 4\theta.$$

So either $3\theta = 4\theta$ or $3\theta = \pi - 4\theta$. The first case means $\theta = 0$; not possible. Thus we have the second case, $7\theta = \pi$, then $\theta = \dfrac{\pi}{7}$.

Comparing with $\theta = \dfrac{\pi}{n}$, clearly $n = 7$.

Answer: 7

Problem 16 Solution
Let p be a prime factor of $2n - 1$, then $p \mid (n - 3)$ or $p \mid (n + 2)$. If $p \mid (n - 3)$, then p is a factor of $(2n - 1) - (n - 3) = n + 2$. Similarly if $p \mid (n + 2)$, then p is a factor of $n - 3$. So p is a common factor of $n - 3$ and $n + 2$, therefore p is a factor of $(n + 2) - (n - 3) = 5$. Hence, $p = 5$. This means that $2n - 1$ is a power of 5.

Let $2n - 1 = 5^t$ where t is a positive integer. Then $n = \dfrac{5^t + 1}{2}$, so

$$\begin{aligned}
\frac{(n - 3)(n + 2)}{2n - 1} &= \frac{\left(\dfrac{5^t - 5}{2}\right)\left(\dfrac{5^t + 5}{2}\right)}{5^t} \\
&= \frac{5^2 (5^{t-1} - 1)(5^{t-1} + 1)}{4 \cdot 5^t}.
\end{aligned}$$

Comparing the exponents of 5, $t \leq 2$. Let $t = 2$, $2n - 1 = 25$, then $n = 13$. Verifying, $\dfrac{(n-3)(n+2)}{2n-1} = \dfrac{10 \times 15}{25} = 6$ is an integer.

Answer: 13

Problem 17 Solution

Since $x > 0$, the expressions are always defined.

Let $y = \sqrt{1+2x}$, so $y > 1$, and

$$4x^2 = (y^2 - 1)^2, \qquad 2x + 9 = y^2 + 8,$$

so the inequality becomes

$$\frac{(y^2-1)^2}{(1-y)^2} \geq y^2 + 8.$$

We know that $y \neq 1$, so

$$\frac{(y^2-1)^2}{(1-y)^2} = \frac{(y+1)^2(y-1)^2}{(y-1)^2} = (y+1)^2,$$

then

$$\begin{aligned}
(y+1)^2 &\geq y^2 + 8, \\
y^2 + 2y + 1 &\geq y^2 + 8, \\
2y &\geq 7, \\
y &\geq 7/2, \\
\sqrt{1+2x} &\geq 7/2, \\
1 + 2x &\geq 49/4, \\
x &\geq 45/8.
\end{aligned}$$

Hence $x \geq \dfrac{45}{8}$ so $P + Q = 45 + 8 = 53$.

Answer: 53

Problem 18 Solution

We want to estimate the upper and lower bounds of $\dfrac{1}{\sqrt{k}}$.

We estimate the lower bound first. Let $k \geq 1$, then

$$\frac{1}{\sqrt{k}} = \frac{2}{2\sqrt{k}} > \frac{2}{\sqrt{k+2}+\sqrt{k}} = \sqrt{k+2}-\sqrt{k}.$$

Letting $k = 1,3,5,7,\ldots,289$ and take the sum,

$$\begin{aligned} S &= 1+\frac{1}{\sqrt{3}}+\frac{1}{\sqrt{5}}+\cdots+\frac{1}{\sqrt{289}} \\ &> (\sqrt{3}-1)+(\sqrt{5}-\sqrt{3})+\cdots+(\sqrt{291}-\sqrt{289}) \\ &= \sqrt{291}-1 \\ &> 16 \end{aligned}$$

Now we estimate the upper bound. Let $k \geq 2$, then

$$\frac{1}{\sqrt{k}} = \frac{2}{2\sqrt{k}} < \frac{2}{\sqrt{k}+\sqrt{k-2}} = \sqrt{k}-\sqrt{k-2}.$$

Letting $k = 3,5,7,\ldots,289$ and take the sum (leave the first term 1 alone),

$$\begin{aligned} S &= 1+\frac{1}{\sqrt{3}}+\frac{1}{\sqrt{5}}+\cdots+\frac{1}{\sqrt{289}} \\ &< 1+(\sqrt{3}-1)+(\sqrt{5}-\sqrt{3})+\cdots+(\sqrt{289}-\sqrt{287}) \\ &= \sqrt{289} \\ &= 17 \end{aligned}$$

Therefore $\lfloor S \rfloor = 16$.

Answer: 16

Problem 19 Solution

We use Principle of Inclusion-Exclusion. There are $\binom{80}{3} = 82160$ integral solutions to

$$a+b+c+d = 100, \quad a \geq 1, b \geq 0, c \geq 2, d \geq 20.$$

Let A be the set of solutions with

$$a \geq 11, b \geq 0, c \geq 2, d \geq 20$$

and B be the set of solutions with

$$a \geq 1, b \geq 0, c \geq 2, d \geq 31.$$

Then $n(A) = \binom{70}{3}$, $n(B) = \binom{69}{3}$, $n(A \cap B) = \binom{59}{3}$ and so

$$n(A \cup B) = \binom{70}{3} + \binom{69}{3} - \binom{59}{3} = 74625.$$

The total number of solutions to

$$a + b + c + d = 100$$

with

$$1 \leq a \leq 10, \; b \geq 0, \; c \geq 2, 20 \leq d \leq 30$$

is thus

$$\binom{80}{3} - \binom{70}{3} - \binom{69}{3} + \binom{59}{3} = 7535.$$

Answer: 7535

Problem 20 Solution

By Cauchy-Schwarz Inequality, Add 1, 4, and 5 to the three

fractions, respectively,

$$
\begin{aligned}
&\frac{a}{b+c}+\frac{4b}{c+a}+\frac{5c}{a+b}+10 \\
=\ &\frac{a}{b+c}+1+\frac{4b}{c+a}+4+\frac{5c}{a+b}+5 \\
=\ &\frac{a+b+c}{b+c}+\frac{4(a+b+c)}{c+a}+\frac{5(a+b+c)}{a+b} \\
=\ &(a+b+c)\left(\frac{1}{b+c}+\frac{4}{c+a}+\frac{5}{a+b}\right) \\
=\ &\frac{1}{2}((b+c)+(c+a)+(a+b))\left(\frac{1}{b+c}+\frac{4}{c+a}+\frac{5}{a+b}\right) \\
\geq\ &\frac{1}{2}(1+2+\sqrt{5})^2 \quad \text{(Cauchy-Schwarz Inequality)} \\
=\ &7+3\sqrt{5},
\end{aligned}
$$

therefore

$$
\frac{a}{b+c}+\frac{4b}{c+a}+\frac{5c}{a+b} \geq 3\sqrt{5}-3,
$$

and equality occurs when $b+c=\dfrac{c+a}{2}=\dfrac{a+b}{\sqrt{5}}$, which means $(a,b,c)=((5+\sqrt{5})k,(5-\sqrt{5})k,(3\sqrt{5}-5)k)$ for any $k>0$.

Since $6<\sqrt{45}=3\sqrt{5}<7$, $3<3\sqrt{5}-3<4$, so $L=4$.

Answer: 4

2.3 ZIML December 2017 Varsity Division

Below are the solutions from the Varsity Division ZIML Competition held in December 2017.
The problems from the contest are available on p.29.

Problem 1 Solution
Draw the diagonal $\overline{AM}$ of the big square, connect $\overline{BM}$.

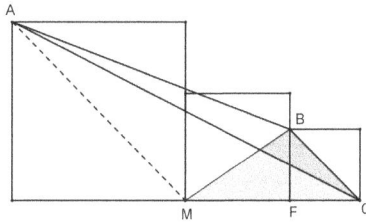

It is clear that $\overline{AM} \parallel \overline{BC}$, thus $[ABC] = [MBC]$. Since $[MBC] = \frac{1}{2}MC \cdot FB = \frac{1}{2} \times 20 \times 8 = 80$, the final answer is 80.

Answer: 80

Problem 2 Solution
First choose two of the teachers to work together, which can be done in $\binom{4}{2}$ ways. This forms 2 distinct groups of teachers (one of size 2 and the others of size 1). Assigning the groups to classes can be done in 3! ways. Hence our final answer is $\binom{4}{2} \cdot 3! = 36$.

Answer: 36

Problem 3 Solution
First, note that $x^2 \geq 0$ implies that $\lfloor 3x \rfloor \geq 0$ and therefore that $x \geq 0$. Furthermore, $3r < x$ implies that $3 \cdot 3r < 3xr < x^2$. It

follows that any solution must satisfy $0 \leq x \leq 3$. From this it is clear that 0 and 3 are solutions as given.

To find the others, note that x^2 must be an integer k satisfying $1 \leq k \leq 8$. By direct computation, we see that the only solutions to $\lfloor 3\sqrt{k} \rfloor = k$ in this range are $\lfloor 3\sqrt{7} \rfloor = 7$ and $\lfloor 3\sqrt{8} \rfloor = 8$. The smaller is $\sqrt{7}$ so $R = 7$.

Answer: 7

Problem 4 Solution
On the first pass, Carrie erases all the odd numbers, leaving only the even numbers (multiples of 2): $2, 4, 6, 8, \ldots, 500$. After the second pass, only the multiples of 4 remain: $4, 8, 12, \ldots, 500$. In general, after the nth pass, only the multiples of 2^n remain. Hence, after the 8th pass, only 256 will remain.

Answer: 256

Problem 5 Solution
The total number of to put the balls in the boxes is 10^5. For the last box to contain exactly 3 balls, first choose which 3 balls go in that box, a total of $\binom{5}{3}$ ways. The other 2 balls can go in any of the remaining 9 boxes, a total of 9^2 ways. Hence the probability is

$$\frac{\binom{5}{3} \cdot 9^2}{10^5} = \frac{10 \cdot 81}{10^5} = \frac{81}{10000}.$$

Thus $P + Q = 10081$.

Answer: 10081

Problem 6 Solution

Using the identity $x^{\log_a y} = y^{\log_a x}$,

$$8^{\log_6(x^2-8x+16)} = 8^{\log_6 4},$$

therefore

$$\log_6(x^2 - 8x + 16) = \log_6 4,$$

and then

$$x^2 - 8x + 16 = 4,$$

which is

$$x^2 - 8x + 12 = 0.$$

Solving this quadratic equation, $x = 2$ or $x = 6$, so the difference is 4.

Answer: 4

Problem 7 Solution

Denote the numbers written on the faces a, b, c, d, e, f, where a and b are on opposite faces, c and d are on opposite faces, and so are e and f. Then we have

$$ace + bce + acf + bcf + ade + bde + adf + bdf = 2015.$$

Factoring,

$$(a+b)(c+d)(e+f) = 2015 = 5 \times 13 \times 31.$$

So

$$a+b+c+d+e+f = 5 + 13 + 31 = 49.$$

Answer: 49

Problem 8 Solution

A and B must be disjoint. Consider cases based on the size of $A \cup B$. Since all the elements in A are larger than the elements in

B, if $A \cup B$ has size k, we have $k - 1$ choices for where to divide into A and B. As there are $\binom{7}{k}$ ways to choose $A \cup B$ of size k, we get a final answer of

$$\sum_{k=2}^{7} (k-1) \cdot \binom{7}{k} = 21 + 2 \cdot 35 + 3 \cdot 35 + 4 \cdot 21 + 5 \cdot 7 + 6 \cdot 1 = 321.$$

Answer: 321

Problem 9 Solution

$A = 0$ is obviously a good value.

If $A > 0$, the two equations $x^2 - 5x - A = 0$ and $x^2 - 5x + A = 0$ combined have two distinct roots. For $A > 0$ the first equation has two real roots, so the second equation must have no real roots. Therefore, $5^2 - 4A < 0$, which leads to $A > \dfrac{25}{4} = 6.25$. Hence the equation has exactly two distinct real roots when $A = 0$ or $A = 7, 8, \ldots, 100$. So there are 95 integers in total.

Answer: 95

Problem 10 Solution

Note that the centers of the four spheres form a regular tetrahedron with side length 20 inches. Hence the height of this tetrahedron is

$$\frac{1}{3} \cdot \sqrt{6} \cdot 20 = \frac{20}{3}\sqrt{6}.$$

Hence the total height is

$$10 + \frac{20}{3}\sqrt{6} + 10 = 20 + \frac{20}{3}\sqrt{6}.$$

Note $2.4 < \sqrt{6} < 2.5$, so

$$16 < \frac{20}{3}\sqrt{6} < 16.\overline{6}.$$

Hence $\lfloor H \rfloor = 36$.

Answer: 36

Problem 11 Solution
First we get $n \geq 4$, otherwise 2 must be adjacent to 1 or 3. With no restrictions there are $n!$ numbers. Now use PIE (Principle of Inclusion-Exclusion). If 2 and 1 are adjacent, bundle them together, $(n-1)!$ ways, and since 2 and 1 can be switched, there are a total of $2 \cdot (n-1)!$ ways. If 2 and 3 are adjacent, we similarly have $2 \cdot (n-1)!$ ways. If 2 is adjacent to both 1 and 3, there are $2 \cdot (n-2)!$ ways. Hence using PIE we have,

$$n! - 2 \cdot 2 \cdot (n-1)! + 2 \cdot (n-2)! = 2400.$$

Simplifying, $(n-2)!(n-2)(n-3) = 2400$, and we can find $n = 7$ by trial and error.

Answer: 7

Problem 12 Solution
First note $\sin C = \dfrac{5}{13}$ as $(5, 12, 13)$ is a Pythagorean Triple. Similarly, $\sin B = \dfrac{7}{25}$. Recall angles in a triangle add up to π radians. Thus we have

$$\sin A = \sin(\pi - B - C) = \sin(B + C)$$
$$= \sin B \cos C + \cos B \sin B = \frac{24}{25} \cdot \frac{5}{13} + \frac{7}{25} \cdot \frac{12}{13}$$
$$= \frac{204}{325}.$$

Hence $Q - P = 121$.

Answer: 121

Problem 13 Solution

Grouping the outside two terms and the inside two terms:

$$(x^2 + 8x + 7)(x^2 + 8x + 15) + 15 = 0.$$

Substituting $y = x^2 + 8x + 7$ and factor in terms of y, we obtain

$$y^2 + 8y + 15 = (y + 5)(y + 3) = 0.$$

Rewriting in terms of x and further factoring gives

$$(x + 2)(x + 6)(x^2 + 8x + 10) = 0.$$

Hence $x = -2, x = -6$ are the only integer solutions, with sum $(-2) + (-6) = -8$.

Answer: -8

Problem 14 Solution

If (a, b, c) is a solution, then (ka, kb, kc) is also a solution for any positive integer k.

So assume $\gcd(a, b, c) = 1$ with $1 \le a \le b \le c$. Since $c \mid (a - b)$, it means $a = b$, and thus $a \mid c$ and so $a = 1$. Therefore $a = b = 1$. Hence all solutions are of the form (k, k, kc) where k and c are positive integers.

Since we have $1 \le A \le B \le C \le 5$ that $c = 1, 2, 3, 4, 5$. This gives, respectively, $5, 2, 1, 1, 1$ triples, so there are 10 in total.

Answer: 10

Problem 15 Solution

By Wilson's Theorem, $31 \mid 30! + 1$. Now

$$30! + 1 = 30 \cdot 29! + 1 \equiv (-1) \cdot 29! + 1 \equiv -(29! - 1) \quad (\text{mod } 31).$$

Hence $31 \mid 29! - 1$. If $p < 31$ is prime, then $p \mid 29!$, so $p \nmid 29! - 1$, hence 31 is the smallest prime factor of $29! - 1$.

Answer: 31

Problem 16 Solution

Since $x > 2$ and $a = \lfloor \log_2 x \rfloor$ is an integer, we get $a \geq 1$. Also $|1-a| \leq 1$ is an integer, so $|1-a| = 0$ or 1. If $|1-a| = 0$, then $a = 1$, and $\sqrt{c-4} = 1$ implies that $c = 5$. Thus $a+c = 6$. If $|1-a| = 1$, then $a = 2$, and $\sqrt{c-4} = 0$ implies that $c = 4$, so $a+c = 6$. We see that in both cases $a+c = 6$.

To find xy, we calculate

$$\log_2 xy = \log_2 x + \log_2 y = a+b+c+d = 6+1 = 7,$$

Hence $xy = 2^7 = 128$.

Answer: 128

Problem 17 Solution

There are $\binom{30}{2}$ ways to pick 2 of the marbles. Factoring, we have $n^2 - 20n + 103 = (n-10)^2 + 3$, so by symmetry, there are 9 pairs of labels, $(9,11),(8,12),\ldots,(1,19)$, that correspond to marbles that weigh the same. Hence the probability is

$$9 \div \binom{30}{2} = \frac{18}{30 \cdot 29} = \frac{3}{145},$$

so $P+Q = 3+145 = 148$.

Answer: 148

Problem 18 Solution

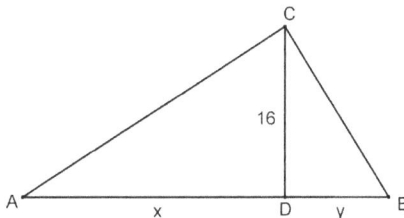

As shown in the diagram, we know $AC + BC = 63$, $CD = 16$. Let $AD = x$, $BD = y$, then by Pythagorean Theorem,

$$AC^2 + BC^2 = (x+y)^2.$$

Also using the area of $\triangle ABC$,

$$[ABC] = \frac{1}{2}(x+y) \cdot 16 = \frac{1}{2}AC \cdot BC,$$

we get

$$AC \cdot BC = (x+y) \cdot 16.$$

Thus

$$
\begin{aligned}
63^2 &= (AC + BC)^2 \\
&= AC^2 + BC^2 + 2AC \cdot BC \\
&= (x+y)^2 + 2(x+y) \cdot 16,
\end{aligned}
$$

therefore

$$
\begin{aligned}
&(x+y)^2 + 32(x+y) - 63^2 \\
&= ((x+y) - 49)((x+y) + 81) \\
&= 0,
\end{aligned}
$$

and

$$x+y = 49 \text{ or } x+y = -81.$$

Taking the positive value, the hypotenuse $x+y = 49$.

Answer: 49

Problem 19 Solution

Reflect point B about $\overrightarrow{OQ}$ to point B', and reflect point A about $\overrightarrow{OP}$ to point A', as shown.

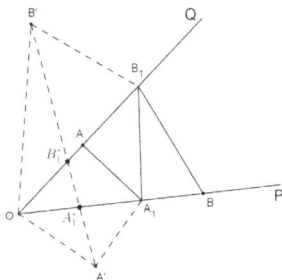

Connect $\overline{A'B'}$ to intersect $\overline{OB}$ at A_1^*, and intersect $\overline{AQ}$ at B_1^*. The points A_1^* and B_1^* are the points that minimize k.

To prove the above conclusion, connect $\overline{A'A_1}$ and $\overline{B'B_1}$. Because of the reflections, for any arbitrary A_1 and B_1 selected, we have $BB_1 = B'B_1$, and $AA_1 = A'A_1$, thus

$$
\begin{aligned}
k &= AA_1 + A_1B_1 + B_1B \\
&= A'A_1 + A_1B_1 + B_1B' \\
&\geq A'B' \\
&= A'A_1^* + A_1^*B_1^* + B_1^*B',
\end{aligned}
$$

where equality occurs when A_1 and B_1 coincide with A_1^* and B_1^* respectively.

To calculate the minimal value, connect $\overline{OA'}$ and $\overline{OB'}$, then $OA' = 3$, $OB' = 4$, and $\angle A'OB' = 3 \times 40° = 120°$. By Law of Cosines,

$$
k_{\min} = A'B' = \sqrt{3^2 + 4^2 - 2 \times 3 \times 4 \cos 120°} = \sqrt{37}.
$$

Therefore $N = 37$.

Answer: 37

Problem 20 Solution

Let $f(x) = x^{1234} + x^{2341} + x^{3412} + x^{4123}$. Let ε be any of the imaginary 5th roots of unity, then $\varepsilon^5 = 1$ and

$$
1 + \varepsilon + \varepsilon^2 + \varepsilon^3 + \varepsilon^4 = 0,
$$

therefore

$$\begin{aligned} f(\varepsilon) &= \varepsilon^{1234} + \varepsilon^{2341} + \varepsilon^{3412} + \varepsilon^{4123} \\ &= \varepsilon^4 + \varepsilon + \varepsilon^2 + \varepsilon^3 \\ &= -1, \end{aligned}$$

so $f(x) + 1$ is divisible by $x^4 + x^3 + x^2 + x + 1$, which means the answer is -1.

Answer: -1

2.4 ZIML January 2018 Varsity Division

Below are the solutions from the Varsity Division ZIML Competition held in January 2018.
The problems from the contest are available on p.35.

Problem 1 Solution

We only need to find the last two digits of 18^{2018}. The last two digits of the powers of 18 follow a pattern:

$$18, 24, 32, 76, 68, 24, 32, 76, 68, \ldots.$$

Starting from 18^2, the pattern repeats every 4 terms. The exponent 2018 has remainder 2 when divided by 4, therefore 18^{2018} has the same last two digits as $18^2 = 324$, so the answer is 24.

Answer: 24

Problem 2 Solution

Multiplying the two equations,

$$\sqrt{(12-y)(4-y)} = y,$$

then

$$48 - 16y + y^2 = y^2,$$

so $y = 3$. Now add the two equations,

$$\sqrt{9} + \sqrt{1} = 2\sqrt{x},$$

and then $x = 4$. Therefore $x^2 + y^2 = 25$.

Answer: 25

Problem 3 Solution

Choose one pair at a time, we get

$$\binom{12}{2}\binom{10}{2}\binom{8}{2}\binom{6}{2}\binom{4}{2}\binom{2}{2} = \frac{12!}{2^6}.$$

However, there should be no ordering among the pairs, so we need to divide by 6!, so the final answer is

$$\frac{12!}{6! \cdot 2^6} = 11 \cdot 9 \cdot 7 \cdot 5 \cdot 3 \cdot 1 = 10395.$$

Answer: 10395

Problem 4 Solution

NOTE: There was originally a typo in the question during the competition. Therefore this question was ignored for grading purposes of the contest.

Extend $\overline{AD}$ and $\overline{BC}$ to intersect at point E. Then $\triangle EAB$ and $\triangle EDC$ are both 30-60-90 triangles. Thus $BE = \sqrt{3}AB = 33$, so $EC = 33 - 15 = 18$, and then $DC = 9$. Since $\angle BCD = 120°$,

$$BD = \sqrt{15^2 + 9^2 - 2 \cdot 15 \cdot 9 \cos 120°} = \sqrt{441} = 21.$$

Answer: 21

Problem 5 Solution

For $1 \le a \le 99$, if $\gcd(a, 100) = 1$, then $\gcd(100 - a, 100) = 1$ as well, and $a \ne 100 - a$. Pair up all the a and $100 - a$ that are relatively prime to 100, and the sum of each pair is 100. There are $\phi(100) = 40$ such numbers, so the number of pairs is $\frac{\phi(100)}{2} = 20$, therefore the total sum is $20 \times 100 = 2000$.

Alternate Solution: The numbers to be added are neither even nor multiples of 5. First add up all the odd numbers less than 100:

$$1 + 3 + 5 + \cdots + 97 + 99 = 2500.$$

Then subtract all the odd multiples of 5:

$$2500 - (5 + 15 + 25 + 35 + 45 + 55 + 65 + 75 + 85 + 95)$$
$$= 2500 - 500 - 2000.$$

Answer: 2000

Problem 6 Solution

$$\frac{10!}{4! \cdot 3! \cdot 2! \cdot 1!} = 12600.$$

Answer: 12600

Problem 7 Solution
The problem gives

$$2836 \equiv 4582 \equiv 5164 \equiv 6522 \pmod{m},$$

so

$$1746 \equiv 582 \equiv 1358 \equiv 0 \pmod{m}.$$

This means m is a common factor of 1746, 582, and 1358. Since

$$\gcd(1746, 582, 1358) = 194,$$

we obtain that $m \mid 194$. So m can be 2, 97, or 194. If $m = 2$, the remainder $r = 0$ which is not a solution. If $m = 97$, $r = 23$. If $m = 194$, $r = 120$. Therefore the maximum possible value of r is 120.

Answer: 120

Problem 8 Solution

Since $\omega^3 = 1$ and $\omega + \omega^2 = -1$, we get

$$\begin{aligned}
&(1-\omega)(1-\omega^2)(1-\omega^4)(1-\omega^8) \\
=\ &(1-\omega)(1-\omega^2)(1-\omega)(1-\omega^2) \\
=\ &(1-\omega-\omega^2+1)^2 \\
=\ &3^2 \\
=\ &9.
\end{aligned}$$

Answer: 9

Problem 9 Solution

Let E and F be the tangent points where the incircle touches the sides $\overline{AC}$ and $\overline{AB}$ respectively, then $BF = BD = \sqrt{3}$ and $CE = CD = 2$. Let r be the inradius, then

$$AE = AF = r\tan\frac{A}{2} = r\tan 30° = \sqrt{3}r.$$

So the semiperimeter

$$s = BD + DC + AE = \sqrt{3} + 2 + \sqrt{3}r.$$

Let $[ABC]$ represent the area of $\triangle ABC$, then according to Heron's formula,

$$\begin{aligned}
&[ABC] \\
=\ &\sqrt{s(s-\sqrt{3}-2)(s-2-\sqrt{3}r)(s-\sqrt{3}-\sqrt{3}r)} \\
=\ &\sqrt{s \cdot \sqrt{3}r \cdot \sqrt{3} \cdot 2} \\
=\ &\sqrt{6rs},
\end{aligned}$$

thus
$$[ABC]^2 = 6rs.$$

We also know that $[ABC] = rs$, therefore

$$[ABC]^2 = 6[ABC],$$

which means
$$[ABC] = 6.$$

Answer: 6

Problem 10 Solution

Choose the 2 boys and 2 girls separately, and arrange the 4 chosen people in the chairs:

$$\binom{8}{2} \cdot \binom{7}{2} \cdot 4! = 14112.$$

Answer: 14112

Problem 11 Solution

Let A be the set of even numbers not exceeding 1000 (multiples of 2), B be the set of multiples of 3, and C be the set of multiples of 5. Then

$$n(A) = 500, \quad n(B) = 333, \quad n(C) = 200.$$

Also, $A \cap B$ is the set of multiples of 6, $B \cap C$ is the set of multiples of 15, $C \cap A$ is the set of multiples of 10, and $A \cap B \cap C$ is the set of multiples of 30, so

$$n(A \cap B) = \left\lfloor \frac{1000}{6} \right\rfloor = 166,$$

$$n(B \cap C) = \left\lfloor \frac{1000}{15} \right\rfloor = 66,$$

$$n(C \cap A) = \frac{1000}{10} = 100,$$

$$n(A \cap B \cap C) = \left\lfloor \frac{1000}{30} \right\rfloor = 33,$$

so

$$n(A \cup B \cup C) = 500 + 333 + 200 - 166 - 66 - 100 + 33$$
$$= 734.$$

Therefore the number of numbers without factors 2, 3 or 5 is

$$1000 - 734 = 266.$$

Answer: 266

Problem 12 Solution

$$\sin 10^\circ \sin 30^\circ \sin 50^\circ \sin 70^\circ$$

$$= \frac{1}{2} \sin 10^\circ \sin 50^\circ \sin 70^\circ$$

$$= \frac{1}{2} \cos 20^\circ \cos 40^\circ \cos 80^\circ$$

$$= \frac{\sin 20^\circ \cos 20^\circ \cos 40^\circ \cos 80^\circ}{2 \sin 20^\circ}$$

$$= \frac{\sin 40^\circ \cos 40^\circ \cos 80^\circ}{4 \sin 20^\circ}$$

$$= \frac{\sin 80^\circ \cos 80^\circ}{8 \sin 20^\circ}$$

$$= \frac{\sin 160^\circ}{16 \sin 20^\circ}$$

$$= \frac{\sin 20^\circ}{16 \sin 20^\circ}$$

$$= \frac{1}{16}.$$

Therefore $\csc 10^\circ \csc 30^\circ \csc 50^\circ \csc 70^\circ = 16$.

Answer: 16

Problem 13 Solution

$$|\sqrt{x-1}-2|+|\sqrt{x-1}-3|=1,$$

so $2 \leq \sqrt{x-1} \leq 3$, thus $5 \leq x \leq 10$.

Answer: 10

Problem 14 Solution

Rearrange the 8 sides of the octagon, so that the side lengths 1 and $\sqrt{2}$ alternate.

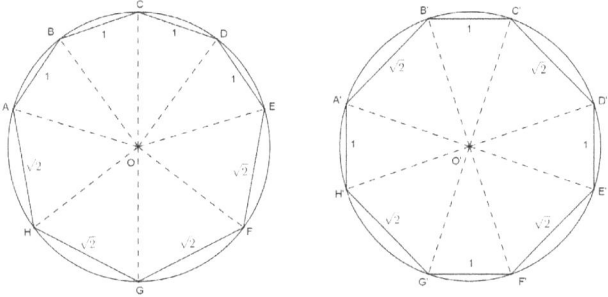

As shown in the diagram, the new octagon $A'B'C'D'E'F'G'H'$ has exactly the same area as the octagon $ABCDEFGH$. Because of symmetry, all the angles of octagon $A'B'C'D'E'F'G'H'$ are equal, and they are all $135°$. To find the area, extend sides $\overline{B'C'}$, $\overline{D'E'}$, $\overline{F'G'}$, and $\overline{H'A'}$ so that they form a square $MNPQ$, as shown below.

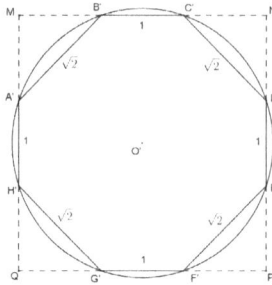

The triangles $\triangle MA'B'$, $\triangle NC'D'$, $\triangle PE'F'$, and $\triangle QG'H'$ are all isosceles right triangles, and so

$$MA' = MB' = NC' = ND' = PE = PF' = QG' = QH' = 1.$$

Thus the length of $\overline{MN}$ is 3, and area of square $MNPQ = 3^2 = 9$, and the combined area of the four isosceles right triangles at the corners is $4 \cdot 1 \cdot 1/2 = 2$. Therefore the area of the octagon $A'B'C'D'E'F'G'H'$ is $9 - 2 = 7$, and thus the original octagon $ABCDEFGH$ also has area 7.

Answer: 7

Problem 15 Solution

Testing $n = 1, 2, 3$, we get that $n = 3$ is a solution. For $n \geq 4$, it is either even or odd. If n is even, let $n = 2k$ (where $k \geq 2$), then $\lfloor n^2/4 \rfloor = k^2$ is not a prime. If n is odd, let $n = 2k + 1$ (also $k \geq 2$), then $\left\lfloor \dfrac{n^2}{4} \right\rfloor = \left\lfloor k(k+1) + \dfrac{1}{4} \right\rfloor = k(k+1)$, also not a prime. Therefore the only solution is $n = 3$.

Answer: 3

Problem 16 Solution

There are two cases: (1) All 3 selected numbers have the same remainder when divided by 3. There are $\dbinom{5}{3} = 10$ for each of the remainders $0, 1, 2$, so in total there are 30 ways. (2) The 3 selected numbers cover the remainders $0, 1, 2$ when divided by 3. There are 5 possibilities for each of the remainders, so the number of choices is $5^3 = 125$. Therefore the total number of ways is $30 + 125 = 155$.

Answer: 155

Problem 17 Solution

In $\triangle ABC$, $\overline{AD}$, $\overline{BE}$, and $\overline{CF}$ are concurrent at O, so by Ceva's Theorem,

$$\frac{BD}{DC} \cdot \frac{CE}{EA} \cdot \frac{AF}{FB} = 1,$$

Since $\dfrac{BD}{DC} = \dfrac{1}{3}$, $\dfrac{CE}{EA} = 1$, we get $\dfrac{AF}{FB} = 3$. Thus $\dfrac{AF}{AB} = \dfrac{3}{4}$.

Considering line $\overline{BOE}$ and $\triangle ADC$, we apply Menelaus' Theorem, then

$$\frac{DB}{BC} \cdot \frac{CE}{EA} \cdot \frac{AO}{OD} = 1;$$

also we know that $\dfrac{DB}{BC} = \dfrac{1}{4}$, thus $\dfrac{AO}{OD} = 4$. Hence, $\dfrac{AO}{AD} = \dfrac{4}{5}$, and consequently

$$\frac{[AFO]}{[ABD]} = \frac{AF \cdot AO \cdot \sin \angle BAD}{AB \cdot AD \cdot \sin \angle BAD} = \frac{3}{4} \cdot \frac{4}{5} = \frac{3}{5},$$

and therefore $\dfrac{[BDOF]}{[ABD]} = \dfrac{2}{5}$. Also, $\dfrac{[ABD]}{[ABC]} = \dfrac{1}{4}$, thus

$$\frac{[BDOF]}{[ABC]} = \frac{2}{5} \cdot \frac{1}{4} = \frac{1}{10}.$$

Hence the ratio

$$\frac{[ABC]}{[BDOF]} = 10.$$

Answer: 10

Problem 18 Solution

Let $u = \sqrt[3]{x-1}, v = \sqrt[3]{y+2}$, then $u^3 + v^3 = -19$, and $u+v = -1$. So $(u+v)(u^2 - uv + v^2) = -19$, and then $u^2 - uv + v^2 = 19$, which means $(u+v)^2 - 3uv = 19$. This gives $uv = -6$. Using Vieta's Theorem, $(u,v) = (-3,2)$ or $(2,-3)$. So $(x,y) = (-26,6)$ or $(9,-29)$. Finally, the possible values of xy are -156 and -261, so the sum of all possible values of xy is -417.

Answer: -417

Problem 19 Solution

Since $792 = 8 \times 9 \times 11$, so we check each separately. $8 \mid \overline{45z}$, thus $z = 6$. Also we get $y = 0$ and $x = 8$ from divisibility rules of 9 and 11. Finally, $100x + 10y + z = 806$.

Answer: 806

Problem 20 Solution

Take reciprocals in the first equation,

$$x_1 + \frac{1}{x_1} = \cdots = x_n + \frac{1}{x_n},$$

thus from the second equation, $n\left(x_1 + \frac{1}{x_1}\right) = \frac{10}{3}$, that is,

$$nx_1^2 - \frac{10}{3}x_1 + n = 0.$$

Since x_1 is a real number, the discriminant $\left(-\frac{10}{3}\right)^2 - 4n^2 \geq 0$, so $n \leq \frac{5}{3}$. But n is a positive integer, thus $n = 1$. Thus, $x_1 = 3$ or $\frac{1}{3}$, therefore by definition of y_1, $y_1 = \log_3 x_1 = 1$ or -1. Hence $|y_1| = 1$. Since $n = 1$, the final sum is 1.

Answer: 1

2.5   ZIML February 2018 Varsity Division

Below are the solutions from the Varsity Division ZIML Competition held in February 2018.
The problems from the contest are available on p.41.

Problem 1 Solution

We are looking for Pythagorean triples. Recall all such triples can
be written as $(a,b,c) = (m^2 - n^2, 2mn, m^2 + n^2)$ for integers m, n.
The number 29 can be a or c. (It cannot be b since b is even.) If
$a = m^2 - n^2 = 29$, then $(m+n)(m-n) = 29$. Since 29 is prime,
we have $m+n = 29$ and $m-n = 1$. Thus $m = 15, n = 14$, and
we get $a = 29, b = 420, c = 421$. If $c = m^2 + n^2 = 29$, the only
possibility is $m = 5, n = 2$. Thus $a = 21, b = 20, c = 29$.

Hence the two triples are $(20, 21, 29)$ and $(29, 420, 421)$. The
sum of the areas of these triangles is

$$\frac{20 \cdot 21}{2} + \frac{29 \cdot 420}{2} = 10 \cdot 21 + 29 \cdot 210 = 6300.$$

Answer: 6300

Problem 2 Solution

Taking the $\log_{30}$ of both sides, we have

$$\log_{30} \sqrt{30} + \log_{30} x^{\log_{30} x} = \log_{30} x^2$$
$$\frac{1}{2} + (\log_{30} x)^2 = 2\log_{30} x$$

The above equation is quadratic, so there are two roots. Let r_1
and r_2 be the two roots of the equation. Let $y = \log_{30} x$, so that
$x = 30^y$. The above equation becomes

$$\frac{1}{2} + y^2 = 2y$$
$$2y^2 - 4y + 1 = 0$$

Let y_1 and y_2 be the two roots of this equation, so that $y_1 + y_2 = \dfrac{4}{2} = 2$. We want

$$r_1 r_2 = 30^{y_1} 30^{y_2} = 30^{y_1+y_2} = 30^2.$$

The answer is $30^2 = 900$.

Answer: 900

Problem 3 Solution

Connect $\overline{AC}$ giving the diagram below.

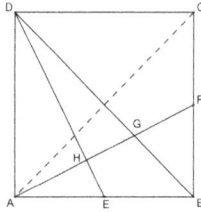

It is clear that G is the centroid of $\triangle ABC$, so the area of $\triangle AGB$ is $\dfrac{1}{3} \cdot \dfrac{1}{2} = \dfrac{1}{6}$ of the full square. Then note $\triangle AHE \sim \triangle ABF$ with a ratio of sides $1 : \sqrt{5}$ hence the area of $\triangle AHE$ is $\dfrac{1}{5} \cdot \dfrac{1}{4} = \dfrac{1}{20}$ of the full square.

This gives that the area of quadrilateral $BEHG$ is

$$\frac{1}{6} - \frac{1}{20} \approx 16.66\% - 5\% \approx 11.67\%$$

and hence $P = 11.7$ when rounded to the nearest tenth.

Answer: 11.7

Problem 4 Solution

By Fermat's Little Theorem, $n^7 \equiv n$ (mod 7) for all positive integers n, so

$$n^2 \equiv n^8 \equiv n^{14} \equiv n^{2+6k} \quad \text{(mod 7)}$$

for all $k = 1, 2, 3, \ldots$. Since $2018 = 2 + 6 \times 336$, we get

$$n^{2018} \equiv n^2 \quad \text{(mod 7) for all } n.$$

Thus

$$1^{2018} + 2^{2018} + 3^{2018} + \cdots + 2018^{2018}$$

$$\equiv \quad 1^2 + 2^2 + 3^2 + \cdots + 2018^2$$

$$\equiv \quad \frac{2018 \cdot 2019 \cdot (2 \cdot 2018 + 1)}{6}$$

$$\equiv \quad 1009 \cdot 673 \cdot 4037$$

$$\equiv \quad 1 \cdot 1 \cdot 5$$

$$\equiv \quad 5 \quad \text{(mod 7)}.$$

Answer: 5

Problem 5 Solution

Label the equilateral triangular regions A, B, C, D, E, F in clockwise order. Consider the colors of A, C, E.

Case (1): A, C, E have the same color. There are $5 \times 4^3 = 320$ ways.

Case (2): Two colors are used to paint A, C, E. There are 3 ways to select the two among A, C, E to be of the same color, say A and C, and 5 ways to choose that color for them. Then 4 ways to choose the one other color for the single region E. Next, region B has 4 choices, and D and F have 3 choices each. In total, $3 \times 5 \times 4 \times 4 \times 3 \times 3 = 2160$ ways.

Case (3): A, C, E are painted with 3 different colors. There are $5 \times 4 \times 3 = 60$ ways for A, C, E. Then each of B, D, F has 3 choices, thus the total number of ways is $60 \times 3^3 = 1620$ ways.

Thus the final answer is $320 + 2160 + 1620 = 4100$ ways.

Answer: 4100

Problem 6 Solution

$$x^2 = (\lfloor x \rfloor + \{x\})^2 = \lfloor x \rfloor^2 + 2\lfloor x \rfloor \{x\} + \{x\}^2,$$

thus

$$\lfloor x^2 \rfloor + \{x^2\} = \lfloor x \rfloor^2 + 2\lfloor x \rfloor \{x\} + \{x\}^2,$$

and since $\{x^2\} = \{x\}^2$,

$$\lfloor x^2 \rfloor = \lfloor x \rfloor^2 + 2\lfloor x \rfloor \{x\},$$

therefore $2\lfloor x \rfloor \{x\}$ must be an integer. Let $\lfloor x \rfloor = n$, and let $2\lfloor x \rfloor \{x\} = k$, then

$$\{x\} = \frac{k}{2n}, \quad 0 \le k \le 2n - 1.$$

It is easy to verify that $x = \lfloor x \rfloor + \{x\} = n + \dfrac{k}{2n}$ where $0 \le k \le 2n - 1$ are all solutions. Hence, there are $2n$ solutions where $n \le x < n+1$ for any positive integer n. For $1 \le x < 7$, there are $2 + 4 + 6 + 8 + 10 + 12 = 42$ solutions.

Answer: 42

Problem 7 Solution

NOTE: There was originally a typo in the question during the competition. Therefore this question was ignored for grading purposes of the contest.

Extend $\overline{CD}$ to intersect the circle at G and let M,N be the midpoints of $\overline{BE}$ and $\overline{GD}$ respectively, as shown in the diagram below.

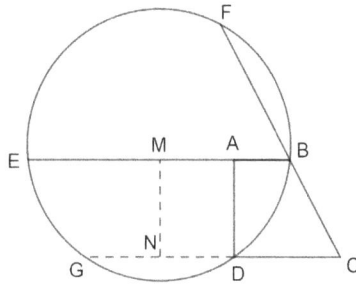

Then $ADNM$ is a rectangle with $AM = DN$. By Power of a Point, $BF = DG$. Therefore

$$BE - BF = BE - DG = 2(BM - AM) = 2AB = 18.$$

Answer: 18

Problem 8 Solution

Construct a triangle ABC such that $AB = 5, BC = 7, CA = 8$. Let P be the point in the interior of $\triangle ABC$ such that

$$\angle APB = \angle BPC = \angle CPA = 120°$$

(this is the first Fermat point of $\triangle ABC$.) Let $x = PA, y = PB, z = PC$, and the given equations are just the results from the Law of Cosines.

To solve for x, y, z, we calculate the area of the triangle. Use $[ABC]$ to represent the area of $\triangle ABC$. The semiperimeter is $(5+7+8)/2 = 10$. By Heron's formula, the area of $\triangle ABC$ is

$$[ABC] = \sqrt{10(10-5)(10-7)(10-8)} = 10\sqrt{3}.$$

Also we calculate $[ABC]$ again by adding up the areas of the three smaller triangles:

$$\begin{aligned} [ABC] &= \frac{1}{2}xy\sin 120° + \frac{1}{2}yz\sin 120° + \frac{1}{2}zx\sin 120° \\ &= \frac{\sqrt{3}}{4}(xy + yz + zx). \end{aligned}$$

Thus we get that $xy + yz + zx = 40$.

Adding the three equations,

$$2(x^2 + y^2 + z^2) + (xy + yz + zx) = 25 + 49 + 64 = 138,$$

hence

$$x^2 + y^2 + z^2 = 49.$$

Then

$$(x+y+z)^2 = x^2 + y^2 + z^2 + 2xy + 2yz + 2zx = 49 + 2 \times 40 = 129,$$

so

$$x + y + z = \sqrt{129}.$$

Subtract the first equation from the second equation,

$$yz + z^2 - x^2 - xy = 24,$$

thus

$$(z+x)(z-x) + y(z-x) = (x+y+z)(z-x) = 24.$$

Therefore $z - x = \dfrac{24}{\sqrt{129}}$. Similarly, subtract the second equation from the third equation,

$$zx + x^2 - y^2 - yz = (x + y + z)(x - y) = 15,$$

so $x - y = \dfrac{15}{\sqrt{129}}$. Then,

$$(x + y + z) + (x - y) - (z - x) = \sqrt{129} + \frac{15}{\sqrt{129}} - \frac{24}{\sqrt{129}},$$

which is

$$3x = \frac{120}{\sqrt{129}},$$

and so

$$x = \frac{40}{\sqrt{129}}.$$

Therefore $m = 40$, $n = 129$, and $m + n = 169$.

Answer: 169

Problem 9 Solution

The least common multiple of $3, 4, 5$ is 60. In the set of numbers $1, 2, \ldots, 60$, there are 20 multiples of 3, 15 multiples of 4, and 5 common multiples of 3 and 4. To remove the multiples of 3 or 4, we use Principle of Inclusion-Exclusion and get

$$60 - 20 - 15 + 5 = 30.$$

However, we need to add back the multiples of 5 that are removed. There are 4 common multiples of 3 and 5, and 3 common multiples of 4 and 5, and 1 common multiples of 3, 4, 5. So the number of remaining numbers is

$$30 + 4 + 3 - 1 = 36.$$

Similarly, among the numbers

$$60k + 1, 60k + 2, \ldots, 60k + 60, \quad (k \in \mathbb{N})$$

36 numbers remain.

Since $2000 = 55 \times 36 + 20$, and the 20th remaining number among $1, 2, \ldots, 60$ is 34, therefore the 200th number is

$$55 \times 60 + 34 = 3334.$$

Answer: 3334

Problem 10 Solution

Let P be the point such that $ABKP$ is a parallelogram. Then it is easy to verify that $EDKP$, $APHC$, and $FPHG$ are all parallelograms. Since $EP = DK$ and $FP = GH$, the triangle EFP is formed by the three lengths EF, GH, KD. This is shown in the diagram below.

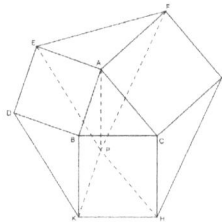

Because of the translation, $\triangle EAP \cong \triangle DBK$ and $\triangle FAP \cong \triangle GCH$, so the area of $\triangle EFP$ is the sum of the areas of $\triangle EAF$, $\triangle DBK$, and $\triangle GCH$.

We know that $AE = AB$, $AF = AC$, and

$$\angle EAF = 360 - 90 - 90 - \angle BAC = 180 - \angle BAC,$$

so the area of $\triangle EAF$ (denoted $[EAF]$) is

$$[EAF] = \frac{1}{2}AE \cdot AF \sin \angle EAF = \frac{1}{2}AB \cdot AC \sin \angle BAC = [ABC].$$

Similarly, $[DBK] = [GCH] = [ABC]$ as well.

Hence the area of the new triangle is 3 times $[ABC]$, which equals 6.

Answer: 6

Problem 11 Solution
Let $u = \sqrt{x-3}$, and $v = \sqrt{x-4}$, then

$$x - 7 = u^2 - 4 \text{ and } x - 5 = v^2 - 1.$$

Thus the equation becomes

$$u - 2 + v - 1 = \sqrt{10},$$

which is

$$u + v = \sqrt{10} + 3.$$

Also we know that $u^2 - v^2 = (x-3) - (x-4) = 1$, therefore

$$u - v = \frac{u^2 - v^2}{u+v} = \frac{1}{\sqrt{10}+3} = \sqrt{10} - 3.$$

so

$$2v = (u+v) - (u-v) = (\sqrt{10}+3) - (\sqrt{10}-3) = 6,$$

and then

$$v = 3,$$

which means

$$\sqrt{x-4} = 3,$$

Therefore

$$x = 13.$$

Thus $x = 13$ is the only solution.

Answer: 13

Problem 12 Solution

If three of these points are collinear, then one of them is the midpoint between the other two.

Case (1): If the two end points are vertices, there are $\binom{8}{2} = 28$;

Case (2): If the two end points are centers of faces, there are 3;

Case (3): If the two end points are midpoints of edges, there are 3 each, and in total $\dfrac{12 \times 3}{2} = 18$.

So the final answer is $28 + 3 + 18 = 49$.

Answer: 49

Problem 13 Solution

From the equation we get

$$x + \sqrt{x^2 - 2018} = \frac{2018}{y + \sqrt{y^2 - 2018}} = y - \sqrt{y^2 - 2018},$$

and

$$y + \sqrt{y^2 - 2018} = \frac{2018}{x + \sqrt{x^2 - 2018}} = x - \sqrt{x^2 - 2018}.$$

Adding these two new equations (add the x side together and y side together), we get $2x = 2y$, so $x = y$, therefore

$$x + \sqrt{x^2 - 2018} = x - \sqrt{x^2 - 2018},$$

which means $\sqrt{x^2 - 2018} = 0$, so $x = y = \pm\sqrt{2018}$. Then

$$
\begin{aligned}
& 2019xy - 2018y^2 + 2017x^2 - 2017^2 \\
= {} & 2019x^2 - 2018x^2 + 2017x^2 - 2017^2 \\
= {} & 2018x^2 - 2017^2 \\
= {} & 2018^2 - 2017^2 \\
= {} & (2018 + 2017)(2018 - 2017) \\
= {} & 4035.
\end{aligned}
$$

Answer: 4035

Problem 14 Solution

Let $x = \angle PBC$, then $\angle ABP = 40° - x$. By the trigonometric form of Ceva's Theorem,

$$
\frac{\sin 80°}{\sin 20°} \cdot \frac{\sin x}{\sin(40° - x)} \cdot \frac{\sin 10°}{\sin 30°} = 1.
$$

Since $\sin 80° \sin 10° = \cos 10° \sin 10° = \dfrac{1}{2}\sin 20°$, and $\sin 30° = \dfrac{1}{2}$, the above equation becomes

$$
\frac{\sin x}{\sin(40° - x)} = 1.
$$

This means $\sin x = \sin(40° - x)$, so either $x = 40° - x$ which implies $x = 20°$, or $180° - x = 40° - x$ which is impossible. Therefore the final answer is $180° - 20° - 30° = 130°$.

Answer: 130

Problem 15 Solution

The following inequalities are true for all real numbers a, b, c, d:

$$2a^2 + \frac{b^2}{2} \geq 2ab,$$

$$\frac{3}{2}b^2 + \frac{3}{2}c^2 \geq 3bc,$$

$$\frac{c^2}{2} + 2d^2 \geq 2cd.$$

Adding the inequalities,

$$2a^2 + 2b^2 + 2c^2 + 2d^2 \geq 2ab + 3bc + 2cd,$$

and equality occurs when $b = c = 2a = 2d$. Therefore the minimum value $M = 2$.

Answer: 2

Problem 16 Solution

The entire volume is 1. Let P and Q be, respectively, the intersections of $\overline{AF}, \overline{CD}$ and $\overline{BF}, \overline{CE}$. Note in fact P and Q are midpoints of all the respective segments. The volume of tetrahedron $F - ABC$ is $\frac{1}{3}$. As P and Q are midpoints, the volume of solid $F - PQC$ is $\frac{1}{4}$ of this volume, hence $\frac{1}{12}$. Therefore the other regions have volumes

$$\frac{1}{3} - \frac{1}{12} = \frac{1}{4}, \frac{1}{3} - \frac{1}{12} = \frac{1}{4}, 1 - 2 \cdot \frac{1}{4} - \frac{1}{12} = \frac{5}{12}.$$

Hence the largest solid has volume $\frac{5}{12}$ with $P + Q = 5 + 12 = 17$.

Answer: 17

Problem 17 Solution

The equation becomes

$$7(x + y) = 3(x^2 - xy + y^2),$$

so
$$3x^2 - 3xy + 3y^2 - 7x - 7y = 0.$$

Treat it as a quadratic equation in x,

$$3x^2 - (3y+7)x + (3y^2 - 7y) = 0.$$

For the equation to have real solutions, the discriminant should be nonnegative:

$$(3y+7)^2 - 4 \cdot 3 \cdot (3y^2 - 7y) \geq 0,$$

so
$$9y^2 + 42y + 49 - 36y^2 + 84y \geq 0,$$

which is
$$27y^2 - 126y - 49 \leq 0.$$

The solution to this quadratic inequality is

$$\frac{21 - 14\sqrt{3}}{9} \leq y \leq \frac{21 + 14\sqrt{3}}{9}$$

The integer values in this interval are $0, 1, 2, 3, 4, 5$. Only $y = 4$ nd $y = 5$ give integer solutions for x, and solve to get $(x, y) = (4, 5)$ or $(5, 4)$. Therefore, the sum of all possible values of y is $4 + 5 = 9$.

Answer: 9

Problem 18 Solution

The plane $x = y$ divides the cube (and any polyhedron it intersects) into two regions, with $x > y$ on one side and $x < y$ on the other side. Similar for the planes $y = z$ and $z = x$. So within each polyhedron, the order of the three coordinates x, y, z are all the same. For example, the set of all points (x, y, z) with $x < y < z$ form one of the polyhedra, and the set of points with $y < x < z$ form another, etc. Therefore, the number of polyhedra is simply the number of ways the three variables x, y, z can be ordered, and thus the answer is $3! = 6$.

Answer: 6

Problem 19 Solution

Use recursion. If there are n points in the triangle, let a_n be the number of small triangles that can be formed. So $a_1 = 3$, and if a new point is added, one of the existing triangles is split into three smaller ones. Thus $a_{n+1} = a_n + 2$ for $n \geq 1$. This is an arithmetic progression, and the general term is $a_n = 3 + 2(n-1) = 2n + 1$. Thus the final answer is $2018 \times 2 + 1 = 4037$.

Answer: 4037

Problem 20 Solution

First recognize $2303 = 47 \cdot 49$. As $\gcd(47, 49) = 1$, we can determine the remainder when divided by 2303 by looking at the remainders when divided by 47 and 49. We first have

$$47^{43} + 49^{46} \equiv 0 + 1 \equiv 1 \pmod{47}$$

using Fermat's Little Theorem. $49 = 7^2$ is not prime, but using Euler's extension to Fermat's Little Theorem (and the fact that $\phi(7^2) = 7^2 - 7 = 42$),

$$47^{43} + 49^{46} \equiv 47 + 0 \equiv 47 \pmod{49}.$$

Since $49 \equiv 2 \pmod{47}$ and $24 \cdot 2 \equiv 1 \pmod{47}$, we have that $47 + 24 \cdot 49 = 1223$ has remainder 1 when divided by 47 and 47 when divided by 49. Therefore 1223 is the remainder when $47^{43} + 49^{46}$ is divided by 2303.

Answer: 1223

2.6 ZIML March 2018 Varsity Division

Below are the solutions from the Varsity Division ZIML Competition held in March 2018.

The problems from the contest are available on p.47.

Problem 1 Solution

There are $4!/2 = 12$ ways to arrange $MMST$, then insert $EEEE$ into the gaps, 5 ways, so the total is $12 \times 5 = 60$ ways. From this, subtract the ways where MM are together: treat MM as the same item, then there are $3! = 6$ ways to arrange $MMST$, and then there is only one way to insert $EEEE$ so that MM is still together. Thus there are 6 ways where $EEEE$ are separate but MM are together. So the final answer is $60 - 6 = 54$ ways.

Answer: 54

Problem 2 Solution

Let
$$S = i + 2i^2 + 3i^3 + \cdots + 99i^{99} + 100i^{100},$$
then
$$iS = i^2 + 2i^3 + 3i^4 + \cdots + 99i^{100} + 100i^{101},$$
so
$$\begin{aligned}(1-i)S &= i + i^2 + i^3 + \cdots + i^{100} - 100i^{101} \\ &= \frac{i(1 - i^{100})}{1-i} + 100i \\ &= -100i,\end{aligned}$$
Thus
$$S = \frac{-100i}{1-i} = -50i(1+i) = 50 - 50i.$$
Thus $a = 50, b = -50$, and $|a| + |b| = 50 + 50 = 100$.

Answer: 100

Problem 3 Solution

Let a be the smallest of the consecutive integers, then

$$a + (a+1) + (a+2) + \cdots + (a+n-1)$$
$$= \frac{n(2a+n-1)}{2} = 6^3 = 216,$$

so

$$n(2a+n-1) = 432.$$

It is easy to see that n and $2a+n-1$ have the opposite parity. In other words, one is odd and the other is even. Also, since $a \geq 1$, $2a+n-1 > n$, therefore $n < \sqrt{432} < 21$.

Since $432 = 2^4 \cdot 3^3$, the possible values for n are 2, 3, 4, 6, 8, 9, 12, 16, 18. Checking the parities of n and $2a+n-1$, they are different when $n = 3$, 9, or 16. So these are the values of n, and their sum is $3 + 9 + 16 = 28$.

Answer: 29

Problem 4 Solution

The trinomial expansion formula is

$$(a+b+c)^n = \sum_{k+m+p=n} \binom{n}{k,m,p} a^k b^m c^p$$

where

$$\binom{n}{k,m,p} = \frac{n!}{k!m!p!}.$$

The constant term of

$$\left(1 + x + \frac{1}{x^2}\right)^{10}$$

is the sum of the coefficients of the terms

$$1^{10-3k} \cdot x^{2k} \cdot \left(\frac{1}{x^2}\right)^k,$$

where $k = 0,1,2,3$, which is

$$\binom{10}{10,0,0} + \binom{10}{7,2,1} + \binom{10}{4,4,2} + \binom{10}{1,6,3} = 4351.$$

Answer: 4351

Problem 5 Solution
Let $x = AB = AC$, so $AE = CD = \dfrac{x}{2}$. Let $y = BC$. Then there are two possibilities: (1) $x + \dfrac{x}{2} = 12, y + \dfrac{x}{2} = 15$, which gives $x = 8, y = 11$. (2) $x + \dfrac{x}{2} = 15, y + \dfrac{x}{2} = 12$, which gives $x = 10, y = 7$. Thus there are two possible values for x and their sum is $8 + 10 = 18$.

Answer: 18

Problem 6 Solution
We first find out what numbers n satisfy $\phi(n) = \dfrac{n}{2}$. Let $p_1, p_2, \ldots$ be the prime factors of n. Since

$$\phi(n) = n\left(1 - \frac{1}{p_1}\right)\left(1 - \frac{1}{p_2}\right)\cdots,$$

we get

$$\left(1 - \frac{1}{p_1}\right)\left(1 - \frac{1}{p_2}\right)\cdots = \frac{1}{2}.$$

The only way for this to happen is that there is only one prime factor, and it is 2. Therefore $n = 2^k$, k is a positive integer.

The largest power of 2 less than 2000 is $2^{10} = 1024$, so the answer is 10.

Answer: 10

Problem 7 Solution

By Vieta's formulas, $\sin\alpha + \sin\beta = \sqrt{2}$ and $\sin\alpha \cdot \sin\beta = \dfrac{c}{2}$.
Therefore,

$$
\begin{aligned}
(\sin\alpha + \sin\beta)^2 &= 2, \\
\sin^2\alpha + 2\sin\alpha\sin\beta + \sin^2\beta &= 2, \\
1 + c &= 2, \\
c &= 1.
\end{aligned}
$$

Also since $\alpha + \beta = 90°$, $\sin\beta = \cos\alpha$, so

$$\sqrt{2} = \sin\alpha + \cos\alpha = \sqrt{2}\sin(\alpha + 45°),$$

thus $\sin(\alpha + 45°) = 1$. Based on the fact that α is an acute angle, $\alpha + 45° = 90°$, then $\alpha = 45°$. Hence, $c = 1$ and $k = 45$, and the final answer is 46.

Answer: 46

Problem 8 Solution

Let the 100 positive integers be $a_1, a_2, \ldots, a_{100}$, and let d be their greatest common divisor. Also let

$$a_1 = db_1, a_2 = db_2, \ldots, a_{100} = db_{100},$$

then $\gcd(b_1, b_2, \ldots, b_{100}) = 1$, and

$$
\begin{aligned}
&\quad a_1 + a_2 + \cdots + a_{100} \\
&= d(b_1 + b_2 + \cdots + b_{100}) \\
&= 101101 \\
&= 101 \times 1001.
\end{aligned}
$$

Hence, $b_1, b_2, \ldots, b_{100}$ cannot all be 1, so

$$b_1 + b_2 + \cdots + b_{100} \geq 1 \times 99 + 2 = 101,$$

therefore
$$d \leq 1001.$$

Now we find a particular example of $a_1, a_2, \ldots, a_{100}$ for which $d = 1001$. Let $a_1 = 2002$ and $a_2 = a_3 = \cdots = a_{100} = 1001$, then their sum equals 101101 and their greatest common divisor is 1001.

Therefore, the largest possible greatest common divisor is 1001.

Answer: 1001

Problem 9 Solution

We find a pattern: If the size of the path is 1×1, there is only 1 way: use a white tile.

If the size of the path is 2×1, there are 2 ways: either use 1 black tile or use 2 white tiles.

If the size of the path is 3×1, there are 3 ways: 1 black and 1 white, 1 white and 1 black, and 3 white tiles.

If the size of the path is 4×1, there are 5 ways: 2 black, 1 black and 2 white, 1 white and 1 black and 1 white, 2 white and 1 black, 4 white.

It is clear that these are the Fibonacci numbers (a more rigorous proof is also available, but we skip it here): $1, 2, 3, 5, 8, 13, 21$. Therefore for the 7×1 path, there are 21 ways.

Answer: 21

Problem 10 Solution

Use the sum-to-product formula and double-angle formula,

$$2\cos\frac{\alpha+\beta}{2}\cos\frac{\alpha-\beta}{2} - 2\cos^2\frac{\alpha+\beta}{2} + 1 = \frac{3}{2},$$

so

$$4\cos\frac{\alpha+\beta}{2}\cos\frac{\alpha-\beta}{2} - 4\cos^2\frac{\alpha+\beta}{2} - 1 = 0,$$

Change signs for all terms, and complete the square,

$$\left(2\cos\frac{\alpha+\beta}{2}-\cos\frac{\alpha-\beta}{2}\right)^2+\sin^2\frac{\alpha-\beta}{2}=0.$$

Thus

$$2\cos\frac{\alpha+\beta}{2}-\cos\frac{\alpha-\beta}{2}=0,$$

and

$$\sin\frac{\alpha-\beta}{2}=0.$$

Since α and β are both acute, $-90°<\frac{\alpha-\beta}{2}<90°$, thus $\alpha-\beta=0$, that is, $\alpha=\beta$, and $\cos\alpha=\frac{1}{2}$, thus $\alpha=60°$. Also we get $\beta=60°$ as well. Therefore, $\alpha+\beta=120°$.

Answer: 120

Problem 11 Solution
Let G be the midpoint of $\overline{BD}$, connect $\overline{EG}$ and $\overline{FG}$, as shown in the diagram on the left.

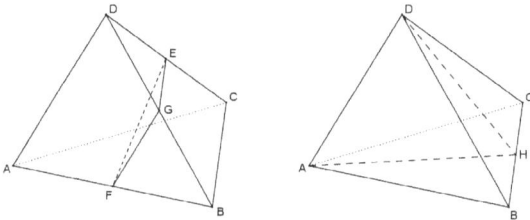

Then $\overline{EG}$ and $\overline{FG}$ are midsegments of $\triangle DBC$ and $\triangle DAB$ respectively, therefore $\overline{EG}\parallel\overline{BC}$ and $\overline{FG}\parallel\overline{AD}$, and the angle between skew lines $\overline{EF}$ and $\overline{DA}$ is the same as the angle $\angle EFG$. In addition, $EG=\frac{1}{2}BC=\frac{1}{2}AD=FG$. In a regular tetrahedron, opposite edges are perpendicular to each other (we shall give a separate proof for this fact), in particular, $\overline{BC} \perp \overline{AD}$, therefore $\overline{EG} \perp \overline{FG}$. Since $EG = FG$, the triangle EFG is an isosceles right triangle, and $\angle EFG = 45°$. Hence, the answer is $45°$.

Note: We now prove that $\overline{BC} \perp \overline{AD}$. Let H be midpoint of $\overline{BC}$, connect $\overline{AH}$ and $\overline{DH}$, as shown in the diagram on the right. Since D-ABC is a regular tetrahedron, $\triangle DBC$ and $\triangle ABC$ are both equilateral triangles, so $\overline{AH} \perp \overline{BC}$ and $\overline{DH} \perp \overline{BC}$. This implies that $\overline{BC}$ is perpendicular to the plane ADH, thus it is perpendicular to all lines in the plane ADH. In particular, $\overline{BC} \perp \overline{AD}$. This completes the proof.

Answer: 45

Problem 12 Solution

Without loss of generality, assume $ar > br > c$. First express $a + c$ and ac in terms of b: $a + c = 2b$, and

$$ac = \frac{1}{2}\left((a+c)^2 - a^2 - c^2\right) = \frac{4b^2 - (84 - b^2)}{2} = \frac{5b^2 - 84}{2}.$$

Since $ar > cr > 0$, a and c are the two distinct positive real roots of the quadratic equation

$$x^2 - 2bx + \frac{5b^2 - 84}{2} = 0.$$

So we have inequalities: $4b^2 - 2(5b^2 - 84)r > 0$, $2br > 0$, and $\frac{5b^2 - 84}{2}r > 0$. These inequalities imply that $\sqrt{\frac{84}{5}}r < br < \sqrt{28}$, and there is only one positive integer in this range, which is $b = 5$.

Answer: 5

Problem 13 Solution

Let $\varepsilon = \cos\dfrac{2\pi}{2n+1} + i\sin\dfrac{2\pi}{2n+1}$ be the imaginary root of unity of degree $2n+1$. Then,

$$1, \varepsilon, \varepsilon^2, \ldots, \varepsilon^{2n}$$

are all the $2n+1$ roots of unity of degree $2n+1$, and

$$z^{2n+1} - 1 = (z-1)(z-\varepsilon)(z-\varepsilon^2)\cdots(z-\varepsilon^{2n}).$$

For the root ε^k ($1 \le k \le n$), match it with its conjugate, which is ε^{2n+1-k}, and we get

$$(z-\varepsilon^k)(z-\varepsilon^{2n+1-k})$$

$$= \left(z - \cos\frac{2k\pi}{2n+1} - i\sin\frac{2k\pi}{2n+1}\right)$$
$$\cdot \left(z - \cos\frac{2k\pi}{2n+1} + i\sin\frac{2k\pi}{2n+1}\right)$$

$$= z^2 - 2z\cos\frac{2k\pi}{2n+1} + 1.$$

Hence in the following factorization,

$$\begin{aligned}
&z^{2n+1} - 1 \\
=\ & (z-1)(z-\varepsilon)(z-\varepsilon^2)\cdots(z-\varepsilon^{2n}) \\
=\ & (z-1)\left((z-\varepsilon)(z-\varepsilon^{2n})\right)\left((z-\varepsilon^2)(z-\varepsilon^{2n-1})\right) \\
& \quad \cdots \left((z-\varepsilon^n)(z-\varepsilon^{n+1})\right) \\
=\ & (z-1)\left(z^2 - 2z\cos\frac{2\pi}{2n+1} + 1\right)\left(z^2 - 2z\cos\frac{4\pi}{2n+1} + 1\right) \\
& \quad \cdots \left(z^2 - 2z\cos\frac{2n\pi}{2n+1} + 1\right) \\
=\ & (z-1)\prod_{k=1}^{n}\left(z^2 - 2z\cos\frac{2k\pi}{2n+1} + 1\right).
\end{aligned}$$

Therefore,

$$\frac{z^{2n+1} - 1}{z-1} = \prod_{k=1}^{n}\left(z^2 - 2z\cos\frac{2k\pi}{2n+1} + 1\right),$$

so

$$z^{2n} + z^{2n-1} + \cdots + z^2 + 1 = \prod_{k=1}^{n} \left(z^2 - 2z\cos\frac{2k\pi}{2n+1} + 1 \right).$$

Let $z = 1$ in the above identity,

$$\begin{aligned}
2n+1 &= \prod_{k=1}^{n} \left(2 - 2\cos\frac{2k\pi}{2n+1} \right) \\
&= \prod_{k=1}^{n} 4\sin^2\frac{k\pi}{2n+1} \\
&= 2^{2n} \prod_{k=1}^{n} \sin^2\frac{k\pi}{2n+1}.
\end{aligned}$$

Therefore

$$\prod_{k=1}^{n} \sin\frac{k\pi}{2n+1} = \frac{\sqrt{2n+1}}{2^n}.$$

Thus, $\dfrac{\sqrt{2n+1}}{2^n} = \dfrac{3}{16}$, and then $n = 4$.

Answer: 4

Problem 14 Solution

There are four cases: (1) $AD = BD = BC$; (2) $AD = BD$, $BC = CD$; (3) $AD = BD = CD$ (this implies $\angle ABC = 90°$, so it is invalid); (4) $AB = BD$ (in this case $\triangle BCD$ cannot be isosceles, so this is not valid either). We only need to analyze the first two cases.

Case (1): $AD = BD = BC$. Let $x = \angle BAC$. Then $\angle ABD = x$. Also $\triangle BCD \sim \triangle ABC$, so $\angle CBD = x$ and then $\angle ACB = \angle ABC = 2x$. Now $x + 2x + 2x = 180°$, thus $x = 36°$.

Case (2): $AD = BD$, $BC = CD$. Let $x = \angle ABC$, then $\angle ABD = x$, and $\angle BDC = 2x$, and so $\angle BDC = 2x$, then $\angle ACB = 180° - 4x$.

Since $AB = AC$, we get $\angle ABC = \angle ACB$, that is $3x = 180° - 4x$, hence $x = \dfrac{180°}{7} = \left(25\dfrac{5}{7}\right)^{\circ}$.

Combining the cases, the sum of all possible degree values of $\angle BAC$ is
$$36 + 25\frac{5}{7} \approx 61.7.$$

Answer: 61.7

Problem 15 Solution
By the Binomial Theorem,

$$(1 - x + x^2)^{19}$$

$$= \left((1-x) + x^2\right)^{19}$$

$$= \binom{19}{0}(1-x)^{19} + \binom{19}{1}(1-x)^{18}x^2 + \binom{19}{2}(1-x)^{17}(x^2)^2 + \cdots,$$

thus
$$a = \binom{19}{0}\binom{19}{1}(-1) = -19,$$

and
$$b = \binom{19}{0}\binom{19}{2} + \binom{19}{1}\binom{18}{0} = 190.$$
Therefore $a + b = 190 - 19 = 171$.

Answer: 171

Problem 16 Solution
Let
$$S = 1 + \frac{1}{\sqrt{2}} + \frac{1}{\sqrt{3}} + \frac{1}{\sqrt{4}} + \cdots + \frac{1}{\sqrt{168}}.$$

Estimate $\dfrac{1}{\sqrt{k}}$ where $1 \le k \le 168$ as follows.

$$\frac{1}{\sqrt{k}} = \frac{2}{2\sqrt{k}} > \frac{2}{\sqrt{k} + \sqrt{k+1}} = 2(\sqrt{k+1} - \sqrt{k}),$$

thus

$$\begin{aligned} S &> 2(\sqrt{2} - \sqrt{1}) + 2(\sqrt{3} - \sqrt{2}) + \cdots + 2(\sqrt{169} - \sqrt{168}) \\ &= 2 \cdot 13 - 290 = 24. \end{aligned}$$

Similarly,

$$\frac{1}{\sqrt{k}} = \frac{2}{2\sqrt{k}} < \frac{2}{\sqrt{k} + \sqrt{k-1}} = 2(\sqrt{k} - \sqrt{k-1}),$$

thus

$$\begin{aligned} S &< 1 + 2(\sqrt{2} - \sqrt{1}) + 2(\sqrt{3} - \sqrt{2}) + \cdots + 2(\sqrt{168} - \sqrt{167}) \\ &= 1 + 2\sqrt{168} - 2 \\ &< 1 + 2 \cdot 13 - 2 \\ &= 25. \end{aligned}$$

Therefore, the largest integer not exceeding S is 24.

Answer: 24

Problem 17 Solution

There are 6 gaps formed by the 5 missed throws. The 4 hits are separated into 3 groups of 2, 1, and 1, and there are

$$\binom{6}{3} \times 3 = 60$$

ways to arrange the groups into the gaps. Therefore the probability is

$$\frac{60}{\binom{9}{4}} = \frac{60}{126} = \frac{10}{21}.$$

Hence $m = 10, n = 21$ and $m + n = 31$.

Answer: 31

Problem 18 Solution

The arc length L of the remaining metal sheet is

$$L = 2\pi \left(1 - \frac{\alpha}{360} \right),$$

and L is also the circumference of the base of the cone. Thus the radius of the base of the cone is

$$r = \frac{L}{2\pi} = 1 - \frac{\alpha}{360},$$

and the height of the cone $h = \sqrt{1 - r^2}$. Then, the volume of the cone is

$$V = \frac{1}{3}\pi r^2 h = \frac{1}{3}\pi r^2 \sqrt{1 - r^2}.$$

In order to maximize V, we define the quantity

$$W = \frac{3V}{\pi} = r^2 \sqrt{1 - r^2},$$

and use the AM-GM inequality to maximize $\dfrac{W^2}{4}$:

$$
\begin{aligned}
\frac{W^2}{4} &= \frac{r^4}{4}(1 - r^2) \\[2mm]
&= \frac{r^2}{2} \cdot \frac{r^2}{2} \cdot (1 - r^2) \leq \left(\frac{\frac{r^2}{2} + \frac{r^2}{2} + 1 - r^2}{3} \right)^3 \\[2mm]
&= \frac{1}{27},
\end{aligned}
$$

where equality occurs when $\dfrac{r^2}{2} = 1 - r^2$, which means

$$r = \sqrt{\frac{2}{3}} = \frac{\sqrt{6}}{3}.$$

Thus, to maximize the volume,

$$1 - \frac{\alpha}{360} = \frac{\sqrt{6}}{3},$$

so

$$\alpha = 360\left(1 - \frac{\sqrt{6}}{3}\right) = 360 - 120\sqrt{6}.$$

Therefore the answer is $360 + 120 + 6 = 486$.

Answer: 486

Problem 19 Solution

Let

$$x = 1999^{1999^{1999^{\cdot^{\cdot^{1999}}}}} = 1999^m,$$

where

$$m = 1999^{1999^{\cdot^{\cdot^{1999}}}}$$

where 1999 appears 1998 times in m. Clearly, m is an odd number. Using the Binomial Theorem,

$$
\begin{aligned}
x &= (2000 - 1)^m \\
&= 2000^m - \binom{m}{1} \cdot 2000^{m-1} + \cdots \\
&\quad - \binom{m}{m-2} \cdot 2000^2 + \binom{m}{m-1} \cdot 2000 - 1 \\
&= K \cdot 10^6 + 2000m - 1,
\end{aligned}
$$

where K is a positive integer. Thus the last 6 digits of x are the same as the last 6 digits of $2000m - 1$.

For any positive integer k, the last 3 digits of 1999^{2k} are 001 and the last 3 digits of 1999^{2k-1} are 999. Since $m = 1999^n$ where n is an odd number, the last 3 digits of m are 999. So the last 6 digits of $2000m - 1$ are the last 6 digits of

$$2000 \times 999 - 1 = 1997999,$$

which are 997999.

Answer: 997999

Problem 20 Solution
Let

$$\frac{y+z+t}{x} = \frac{z+t+x}{y} = \frac{t+x+y}{z} = \frac{x+y+z}{t} = k,$$

then

$$
\begin{aligned}
y+z+t &= kx, \\
z+t+x &= ky, \\
t+x+y &= kz, \\
x+y+z &= kt.
\end{aligned}
$$

Adding them up,

$$3(x+y+z+t) = k(x+y+z+t),$$

so

$$(k-3)(x+y+z+t) = 0.$$

Hence either $k = 3$ or $x+y+z+t = 0$.

Case 1: $k = 3$, thus $y+z+t = 3x$ and $z+t+x = 3y$. Subtracting, $y - x = 3(x-y)$, and then $x = y$. Similarly, $y = z = t = x$. In this case the required expression has value 4.

Case 2: $x+y+z+t = 0$. Thus $x+y = -(z+t)$, and so on, so each term in the required expression is -1, and the final absolute value is still 4.

So there is only one possible value, which is 4.

Answer: 4

2.7 ZIML April 2018 Varsity Division

Below are the solutions from the Varsity Division ZIML Competition held in April 2018.

The problems from the contest are available on p.53.

Problem 1 Solution

Note: The original problem stated $DF : FC = 1 : 3$ instead of $DF : FC = 1 : 1$, in which case the ratio is not constant. This problem was not counted towards the final score of the contest.

As EG and FH are parallel to DB, and E and F are the midpoints of AD and DC, we have

$$[DBG] = [DBE] = \frac{1}{2}[ADB]$$

and

$$[BDH] = [BDF] = \frac{1}{2}[CDB],$$

here $[ABC]$ denotes the area of $\triangle ABC$. Thus,

$$[DBG] + [BDH] = \frac{1}{2}[ABC].$$

Now extend CB to a point G' such that $BG = GG'$ and AB to a point H' such that $BH = HH'$. Then AG' is parallel to CH' and $[G'AH'] = [G'AC]$, thus by removing $[G'AB]$ from both triangles we get $[G'BH'] = [ABC]$, and so $[GBH] = \frac{1}{4}[ABC]$. Hence

$$[DHG] = \frac{3}{4}[ABC],$$

which means $\dfrac{[ABC]}{[DHG]} = \dfrac{4}{3}$ and $P + Q = 7$.

Answer: 7

Problem 2 Solution

Consider the binomial expansion of $(1+x)^{2017}$, and let $x = i$ and $x = -i$ where $i = \sqrt{-1}$, then

$$
\begin{aligned}
N &= \frac{1}{2}\left((1+i)^{2017} + (1-i)^{2017}\right) \\
&= \frac{1}{2}\left(\left(\sqrt{2}e^{i\pi/4}\right)^{2017} + \left(\sqrt{2}e^{-i\pi/4}\right)^{2017}\right) \\
&= \frac{1}{2}\left(2 \cdot 2^{2017/2} \cos\frac{\pi}{4}\right) \\
&= 2^{1008},
\end{aligned}
$$

thus $\log_2 N = 1008$.

Answer: 1008

Problem 3 Solution

First we use the fact that $7^2 \equiv 49 \equiv (-1) \pmod{50}$ to get

$$
\begin{aligned}
(7^{2019} + 46)^{2018} &\equiv (7 \cdot (-1)^{1009} - 4)^{2018} \\
&\equiv (-11)^{2018} \\
&\equiv 11^{2018} \pmod{50}.
\end{aligned}
$$

Then also using the fact that $11^5 \equiv 1 \pmod{50}$ we simplify

$$
\begin{aligned}
11^{2018} &\equiv 11^3 \cdot (11^5)^{403} \\
&\equiv 11^3 \\
&\equiv 1331 \\
&\equiv 31 \pmod{50}
\end{aligned}
$$

so the remainder is 31.

Answer: 31

Problem 4 Solution

By Vieta's Theorem,

$$\begin{aligned} a+b+c &= -a, \\ ab+bc+ca &= b, \\ abc &= -c. \end{aligned}$$

If either $a = 0$ or $b = 0$, we can derive that $a = b = c = 0$. Since a, b, c are not all 0, we must have $a \neq 0$ and $b \neq 0$.

Consider two cases: $c = 0$ and $c \neq 0$.

Case (1): $c = 0$. In this case, $a + b = -a$ and $ab = b$, thus $a = 1$ and $b = -2$. In fact, the roots of the equation

$$x^3 + x - 2x = 0$$

are exactly 1, -2, and 0. So we have solution $a = 1, b = -2, c = 0$, and then $a^2 + b^2 + c^2 = 5$.

Case (2): $c \neq 0$. In this case, $ab = -1$, so $b = -\dfrac{1}{a}$, we have the system of equations

$$\begin{aligned} a - \frac{1}{a} + c &= -a, \\ -1 + c\left(-\frac{1}{a} + a\right) &= -\frac{1}{a}. \end{aligned}$$

Thus $c = \dfrac{1}{a} - 2a$, and

$$-1 + \left(\frac{1}{a} - 2a\right)\left(-\frac{1}{a} + a\right) = -\frac{1}{a}.$$

Simplifying the above equation to get

$$2a^4 - 3a^2 + a = 0.$$

Factoring,

$$a(a-1)(2a^2+2a-1)=0,$$

where the only nonzero rational root is $a = 1$. Hence, $b = -1$ and $c = -1$. In fact, the roots of the equation

$$x^3 + x^2 - x - 1 = 0$$

are exactly $1, -1$, and -1. Therefore, $a = 1$, $b = -1$, and $c = -1$ is also a set of solutions, and $a^2 + b^2 + c^2 = 3$.

Combining the two cases, the sum of all possible values of $a^2 + b^2 + c^2$ is $5 + 3 = 8$.

Answer: 8

Problem 5 Solution

We can write $0.\overline{abcd} = \dfrac{\overline{abcd}}{9999}$ for any repeating decimal with cycle length four. Therefore N is a factor of $\overline{abcd}$ and M is a factor of 9999. To minimize $N + M$ we want the greatest common divisor of $\overline{abcd}$ and 9999 to be as large as possible.

The factors of 9999 are $9999, 1111, 909, 303, 101, 99, \ldots$, so we consider multiples of these as $\overline{abcd}$. Clearly 9999 doesn't work and any multiple of 1111 or 101 gives a fraction of the form $\dfrac{N}{9}$ or $\dfrac{N}{99}$ which has a cycle length of 1 or 2 (and hence a, b, c, d are not distinct).

Thus consider numbers of the form $\dfrac{99k}{9999} = \dfrac{k}{101}$ for integers k. When $k = 1$ we have

$$\frac{99}{9999} = \overline{0099} = 0.00990099\ldots$$

so a, b, c, d are not distinct. However, when $k = 2$ we have

$$\frac{198}{9999} = \overline{0198} = 0.01980198\ldots$$

so a, b, c, d are all distinct as needed. This minimizes $N + M$, with $N + M = 2 + 101 = 103$.

Answer: 103

Problem 6 Solution
Since $AC = AD + DC = 3$, we get

$$\frac{AD}{AB} = \frac{2}{\sqrt{6}} = \frac{\sqrt{6}}{3}, \quad \frac{AB}{AC} = \frac{\sqrt{6}}{3} = \frac{AD}{AB},$$

thus $\triangle ABD \sim \triangle ACB$, so $\angle ABC = \angle ADB = 60°$.

Let E be the point on $\overline{BD}$ such that $\overline{AE} \perp \overline{BD}$, then $\triangle ADE$ is a 30-60-90 triangle, and $AE = \sqrt{3}$. Hence, $AB = \sqrt{6} = AE \cdot \sqrt{2}$, so $\triangle AEB$ is a 45-45-90 triangle, and then $\angle ABD = 45°$.

Therefore, $\angle CBD = \angle ABC - \angle ABD = 60° - 45° = 15°$.

Answer: 15

Problem 7 Solution
Let E be the event that the rolls add up to 6 and H, T the events you get heads, tails. We have

$$P(H) = x, P(T) = 1 - x, P(E|H) = \frac{1}{6}, P(E|T) = \frac{2}{36} = \frac{1}{18}.$$

Therefore

$$0.5 = \frac{P(H) \cdot P(E|H)}{P(H) \cdot P(E|H) + P(T) \cdot P(E|T)}$$

$$= \frac{x \cdot \dfrac{1}{6}}{x \cdot \dfrac{1}{6} + (1 - x) \cdot \dfrac{1}{18}}$$

$$= \frac{3x}{2x + 1}$$

Solving we have $x = \dfrac{1}{4}$. Thus $K = 25$.

Answer: 25

Problem 8 Solution

Let the 3-digit number be $100a + 10b + c$ where $1 \le a \le 9$ and $0 \le b, c \le 9$. We now find the minimum of the function

$$f(a,b,c) = \frac{100a + 10b + c}{a + b + c}.$$

Transform the function,

$$f(a,b,c) = 1 + \frac{99a + 9b}{a + b + c}.$$

For any fixed values of a and b, the fraction $\dfrac{99a + 9b}{a + b + c}$ is mini-mized when $c = 9$, thus

$$f(a,b,c) \ge 1 + \frac{99a + 9b}{a + b + 9} = 10 + \frac{90a - 81}{a + b + 9}.$$

For any fixed value of a, the fraction $\dfrac{90a - 81}{a + b + 9}$ is minimized when $b = 9$, thus

$$f(a,b,c) \ge 10 + \frac{90a - 81}{a + 18} = 100 - \frac{1701}{a + 18}.$$

The last expression $100 - \dfrac{1701}{a + 18}$ is minimized when $a = 1$, thus

$$f(a,b,c) \ge 100 - \frac{1701}{19} = \frac{199}{19},$$

and equality occurs when $a = 1, b = c = 9$. Hence $m = 199$, $n = 19$, and $m + n = 199 + 19 = 218$.

Answer: 218

Problem 9 Solution

Note that $\gcd(40,64) = 8$. Therefore the problem is equivalent to counting numbers ≤ 125 that can be written in the form $r \cdot 5 + s \cdot 8$ for non-negative integers r, s.

The multiples of 8: 8, 16, 24, and 32, are respectively 3, 1, 4, and 2 modulo 5. Hence, for example, any number $\equiv 1 \pmod 5$, starting with 16 will work (as it can be written as $r \cdot 5 + 2 \cdot 8$) so there are $\lfloor 16 \div 5 \rfloor = 3$ that do not work. These results are summarized in the table below:

n (mod 5)	Works for all $n \geq$	Number that do not work
0	0	0
1	16	3
2	32	6
3	8	1
4	24	4

Thus out of the $125 - 0 + 1 = 126$ numbers, 14 do not work. This gives an answer of $126 - 14 = 112$.

Answer: 112

Problem 10 Solution

Connect $\overline{A_1 D}$ and $\overline{B_1 C}$. Clearly, $A_1 B_1 CD$ is a parallelogram, and $\overline{AB}$ is parallel to the plane $A_1 B_1 CD$, which contains the line $\overline{DB_1}$. So the distance between skew lines $\overline{AB}$ and $\overline{DB_1}$ is equal to the distance between line $\overline{AB}$ and the plane $A_1 B_1 CD$, which in turn equals the distance from point B to line $\overline{B_1 C}$, which is an altitude h of $\triangle B_1 BC$. Since $\angle B_1 BC = 90°$, $BC = 4$, $BB_1 = 3$, we get $B_1 C = 5$, and the altitude h is

$$h = \frac{BC \cdot BB_1}{B_1 C} = \frac{4 \times 3}{5} = 2.4,$$

and this is the distance we are looking for.

Answer: 2.4

Problem 11 Solution

From $f(x+1)(1-f(x)) = 1 + f(x)$, we get

$$f(x+1) = \frac{1+f(x)}{1-f(x)},$$

so

$$f(x+2) = \frac{1 + \dfrac{1+f(x)}{1-f(x)}}{1 - \dfrac{1+f(x)}{1-f(x)}} = -\frac{1}{f(x)},$$

and thus

$$f(x+4) = -\frac{1}{f(x+2)} = -\frac{1}{-\dfrac{1}{f(x)}} = f(x).$$

Hence $f(x)$ is a periodic function with period 4. Therefore

$$f(2018) = f(4 \cdot 504 + 2) = f(2) = -\frac{1}{f(0)} = -\frac{1}{5} = -0.2.$$

Answer: -0.2

Problem 12 Solution

$2018 = 2 \cdot 1009$. Therefore we want m, n such that

$$\frac{m}{n} = \frac{2^r \cdot 1009^s}{2^{r'} \cdot 1009^{s'}} \text{ with } \max(r, r') = 1, \max(s, s') = 1,$$

for non-negative integers r, s. In total there are

$$(2^2 - 1^2)(2^2 - 1^2) = 9$$

such numbers. One of these is $\dfrac{2018}{2018} = 1$ and by symmetry the others are half $r > 1$ and half $r < 1$. Therefore $8 \div 2 = 4$ such rational numbers exist.

Answer: 4

Problem 13 Solution

Let $x = 1 + pi$, $y = 1 + qi$, and $z = 1 + ri$ be complex numbers, then $\arctan p$, $\arctan q$, and $\arctan r$ are the arguments of x, y, and z respectively. Define

$$
\begin{aligned}
w &= xyz \\
&= (1 + pi)(1 + qi)(1 + ri) \\
&= (1 - pq - qr - rp) + (p + q + r - pqr)i,
\end{aligned}
$$

then

$$
\arg w = \arctan p + \arctan q + \arctan r = -\frac{\pi}{2},
$$

which means w is in fact a pure imaginary number, so its real part

$$
1 - pq - qr - rp = 0,
$$

therefore $pq + qr + rp = 1$, which is the only possible value.

Answer: 1

Problem 14 Solution

We use $[CAD]$ to represent the area of $\triangle CAD$, and use similar notations for other areas.

Since $\overline{DE}$ is the diameter of the circle, $\angle DCE = 90°$.

Since $AD = DE$, the area ratio

$$
\frac{[CAD]}{[CAE]} = \frac{1}{2},
$$

and also

$$
\frac{[CAD]}{[CAE]} = \frac{\frac{1}{2} CA \cdot CD \sin \alpha}{\frac{1}{2} CA \cdot CE \sin(\alpha + 90°)} = \frac{CD \sin \alpha}{CE \cos \alpha} = \frac{CD}{CE} \tan \alpha,
$$

therefore

$$
\frac{CD}{CE} \tan \alpha = \frac{1}{2}.
$$

Similarly, since $BE = DE$, the area ratio

$$\frac{[CBE]}{[CBD]} = \frac{1}{2},$$

and also

$$\frac{[CBE]}{[CBD]} = \frac{\frac{1}{2}CB \cdot CE \sin\beta}{\frac{1}{2}CB \cdot CD \sin(\beta + 90°)} = \frac{CE \sin\beta}{CD \cos\beta} = \frac{CE}{CD}\tan\beta.$$

therefore

$$\frac{CE}{CD}\tan\beta = \frac{1}{2}.$$

Consequently,

$$\tan\alpha \cdot \tan\beta = \frac{1}{2} \cdot \frac{1}{2} = \frac{1}{4} = 0.25.$$

Answer: 0.25

Problem 15 Solution

One of the primes must be 11, so let the numbers be $11, p, q$ (assume $pr < q$). We have

$$11pq = 11(11 + p + q)$$

Thus

$$pq = 11 + p + q \Rightarrow pq - p - q + 1 = (p-1)(q-1) = 12$$

after completing the rectangle. p, q are positive, so the possibilities are $(p-1, q-1) = (1, 12)$, $(2, 6)$, or $(3, 4)$. Hence $(p, q) = (2, 13)$, $(3, 7)$, or $(4, 5)$. Only the first two are such that p and q are prime, so our possible sets are $2, 11, 13$ or $3, 7, 11$. The ratio of the products is

$$\frac{2 \cdot 11 \cdot 13}{3 \cdot 7 \cdot 11} = \frac{26}{21}$$

and thus $M + N = 47$.

Answer: 47

Problem 16 Solution

Let $y^2 = \sqrt{x}$, then the expression becomes

$$\left(y^2 + \frac{1}{4y^2} - 1\right)^6 = \left(y - \frac{1}{2y}\right)^{12},$$

so the constant term is

$$(-1)^6 \binom{12}{6} \left(\frac{1}{2}\right)^6 = \frac{924}{32} = \frac{231}{16}.$$

As $231 \div 16 = 14R7$, rounded to the nearest integer the answer is 14.

Answer: 14

Problem 17 Solution

Let

$$\begin{aligned} f(x) &= \sqrt{12x - x^2 - 11} + \sqrt{68x - x^2 - 256} \\ &= \sqrt{(x-1)(11-x)} + \sqrt{(x-4)(64-x)}, \end{aligned}$$

so the domain of $f(x)$ is $[1,11] \cap [4,64] = [4,11]$.

Let $a = \sqrt{x-1}$, $b = \sqrt{11-x}$, $c = \sqrt{x-4}$, and $d = \sqrt{64-x}$, then by Cauchy-Schwarz Inequality,

$$\begin{aligned} f(x) = ab + cd &\leq \sqrt{(a^2 + d^2)(b^2 + c^2)} \\ &= \sqrt{(x-1+64-x)(11-x+x-4)} \\ &= \sqrt{63 \cdot 7} \\ &= 21, \end{aligned}$$

where equality occurs when $\dfrac{a}{b} = \dfrac{d}{c}$, that is

$$\frac{x-1}{11-x} = \frac{64-x}{x-4}$$

which means $x = 10$, which is within the domain $[4, 11]$.

So the final answer is 21.

Answer: 21

Problem 18 Solution
We know using Wilson's theorem that $70! \equiv -1 \pmod{71}$. Then note
$$\begin{aligned}
70! &\equiv 61! \cdot 62 \cdot 63 \cdots 70 \\
&\equiv 61! \cdot (-9)(-8)\cdots(-1) \\
&\equiv 61! \cdot (-1)(1 \cdot 2 \cdot 5 \cdot 7)(3 \cdot 4 \cdot 6)(8 \cdot 9) \\
&\equiv 61! \cdot (-1)(-1)(1)(1) \\
&\equiv 61! \pmod{71}
\end{aligned}$$
Thus $61! \equiv 70! \equiv -1 \equiv 70 \pmod{71}$ so the remainder is 70.

Answer: 70

Problem 19 Solution
For a polygon with n sides, there are $\dfrac{n(n-3)}{2}$ total diagonals. Any intersection of these diagonals is determined (uniquely) by 4 vertices of the polygon. Hence there are $\binom{n}{4}$ intersections. Each of these intersections divides a diagonal into 2 additional pieces (since no three diagonals go through the same point). Therefore in total the diagonals are divided into
$$\frac{n(n-3)}{2} + 2 \cdot \binom{n}{4} = \frac{n(n-3)}{12}(6 + (n-1)(n-2))$$
pieces. Some trial and error gives $n = 11$. (One useful observation is that 704 is a multiple of 11.)

Answer: 11

Problem 20 Solution

For each i, apply Stewart Theorem,

$$BC \cdot AP_i^2 + BC \cdot BP_i \cdot CP_i = BP_i \cdot AC^2 + CP_i \cdot AB^2.$$

Since $AB = AC = 5$,

$$BC(AP_i^2 + BP_i \cdot CP_i) = 5^2(BP_i + CP_i) = 25BC,$$

therefore
$$AP_i^2 + BP_i \cdot CP_i = 25.$$

Hence, $k_i = 25$ for $i = 1, 2, \ldots, 40$, and so

$$k_1 + k_2 + \cdots + k_{40} = 25 \times 40 = 1000.$$

Answer: 1000

2.8   ZIML May 2018 Varsity Division

Below are the solutions from the Varsity Division ZIML Competition held in May 2018.

The problems from the contest are available on p.59.

Problem 1 Solution
Between 1 and 800 inclusive, there are 28 perfect squares, 9 perfect cubes, and 3 perfect 6th powers (which are $1, 2^6, 3^6$). Thus the sequence skips

$$28 + 9 - 3 = 34$$

numbers.

From 801 to 834, there are no perfect squares or perfect cubes, so the 800th number in the sequence is 834.

Answer: 834

Problem 2 Solution
Since $2500 \leq \overline{25ab} < 2600$, we get

$$2500 \leq 2^5 \cdot a^b < 2600,$$

thus
$$78 \leq a^b \leq 82.$$

The only integer power between 78 and 82 is 81, so

$$2^5 \cdot a^b = 32 \cdot 81 = 2592,$$

so the only possibility is $a = 9$ and $b = 2$. Thus the final answer is 92.

Answer: 92

Problem 3 Solution

From $\log_2 f(a) = 3$ we get that $f(a) = 8$, so

$$2a^2 - 2a + k = 8;$$

also, $f(\log_2 a) = k$ means that $2\log_2^2 a - 2\log_2 a + k = k$, thus $2(\log_2^2 a - \log_2 a) = 0$, and then either $\log_2 a = 0$ or $\log_2 a = 1$. If $\log_2 a = 0$, then $a = 1$, which doesn't satisfy the requirement that $a \neq 1$. If $\log_2 a = 1$, then $a = 2$, so $k = 4$. Therefore,

$$f(x) = 2x^2 - 2x + 4 = 2\left(x - \frac{1}{2}\right)^2 + \frac{7}{2}.$$

Hence

$$f(\log_2 x) = 2\left(\log_2 x - \frac{1}{2}\right)^2 + \frac{7}{2}.$$

Let $x = \sqrt{2}$, then $\log_2 x = \frac{1}{2}$, and so the minimum value of $f(\log_2 x)$ is $\frac{7}{2} = 3.5$.

Answer: 3.5

Problem 4 Solution

The 12 edges of a cube has 3 directions in total. By symmetry, it is clear that the plane $AB'D'$ intersects the edges at equal angles. So θ equals the angle between AA' and the plane $AB'D'$. Let $AA' = 1$, then $A'C = \sqrt{3}$, and it is easy to verify that $\overline{A'C}$ is perpendicular to plane $AB'D'$, and is trisected by the planes $AB'D'$ and $BC'D'$. Thus

$$\sin\theta = \frac{1}{\sqrt{3}},$$

so $\csc^2\theta = (\sqrt{3})^2 = 3$.

Answer: 3

Problem 5 Solution

$$
\begin{aligned}
& 999999999999 \\
= & 10^{12} - 1 \\
= & (10^6 + 1)(10^6 - 1) \\
= & (10^2 + 1)(10^4 - 10^2 + 1)(10^2 - 1)(10^4 + 10^2 + 1) \\
= & (101)(9901)(99)(10^4 + 2 \cdot 10^2 + 1 - 10^2) \\
= & (101)(9901)(3^2)(11)(101^2 - 10^2) \\
= & (101)(9901)(3^2)(11)(111)(91) \\
= & 3^3 \times 7 \times 11 \times 13 \times 37 \times 101 \times 9901.
\end{aligned}
$$

So the sum of distinct prime factors is

$$
3 + 7 + 11 + 13 + 37 + 101 + 9901 = 10073
$$

Answer: 10073

Problem 6 Solution

The curve is in fact the circle with

$$
(\arcsin k, \arcsin k) \text{ and } (-\arccos k, \arccos k)
$$

as end points of a diameter. Let $\theta = \arcsin k \in \left[-\dfrac{\pi}{2}, \dfrac{\pi}{2}\right]$, then $\arccos k = \dfrac{\pi}{2} - \theta$. So the two endpoints of the diameter are (θ, θ) and $\left(-\dfrac{\pi}{2} + \theta, \dfrac{\pi}{2} - \theta\right)$. Hence the center of the circle is $\left(-\dfrac{\pi}{4} + \dfrac{\theta}{2}, \dfrac{\pi}{4}\right)$.

The line $y = \dfrac{\pi}{4}$ clearly passes through the center of this circle, therefore $\overline{MN}$ is another diameter of the circle, thus the length is

$$
\sqrt{\left(\frac{\pi}{2}\right)^2 + \left(2\theta - \frac{\pi}{2}\right)^2} \geq \frac{\pi}{2}.
$$

So the minimum length is $\dfrac{\pi}{2} \approx 1.57$.

Answer: 1.57

Problem 7 Solution

There are $\dbinom{5}{3} = 10$ ways to choose 3 even numbers, and

$$\binom{5}{1}\binom{5}{2} = 50$$

ways to choose 1 even and 2 odd numbers, so there are 60 ways to choose 3 numbers with even sums.

Among these triples, the following have sums less than 10:

$(0,2,4)$, $(0,2,6)$, $(0,1,3)$, $(0,1,5)$, $(0,1,7)$, $(0,3,5)$, $(2,1,3)$, $(2,1,5)$, $(4,1,3)$.

There 9 such triples in total.

So the final answer is $60 - 9 = 51$.

Answer: 86

Problem 8 Solution

Let $z = x + yi$, where x and y are real numbers, then $\bar{z} = x - yi$. So $\dfrac{z}{20} = \dfrac{x}{20} + \dfrac{y}{20}i$. According to the question, $0 \le \dfrac{x}{20} \le 1$ and $0 \le \dfrac{y}{20} \le 1$, therefore $0 \le x \le 20$ and $0 \le y \le 20$. This is a square with side length 20. Also

$$\frac{20}{\bar{z}} = \frac{20}{x - yi} = \frac{20x}{x^2 + y^2} + \frac{20y}{x^2 + y^2}i,$$

thus

$$0 \le \frac{20x}{x^2 + y^2} \le 1, \quad 0 \le \frac{20y}{x^2 + y^2} \le 1,$$

and these lead to the inequalities

$$x \geq 0, \quad y \geq 0, \quad (x-10)^2 + y^2 \geq 100, \quad x^2 + (y-10)^2 \geq 100.$$

So the point z is inside the square with side length 20 and outside the two semicircles with radii 10, as shown.

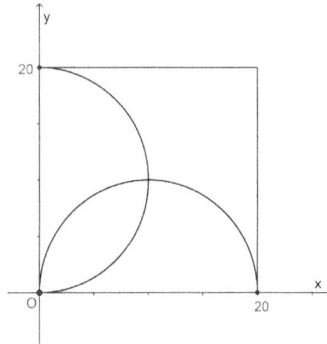

The area of overlap of the two semicircles is

$$10^2 \pi / 2 - 10^2 = 50(\pi - 2),$$

so the region in the square and outside the two semicircles is

$$
\begin{aligned}
20^2 - 10^2 \pi + 50(\pi - 2) &= 300 - 50\pi \\
&\approx 300 - 50 \times 3.14 \\
&= 143.
\end{aligned}
$$

Answer: 143

Problem 9 Solution
Complete the square, and apply the Sum-to-Product and Product-

to-Sum formulas,

$$\sin^2 80° + \sin^2 40° - \cos 50° \cos 10°$$

$$= (\sin 80° - \sin 40°)^2 + \cos 50° \cos 10°$$

$$= 2^2 \sin^2 20° \cos^2 60° + \frac{1}{2}(\cos 60° + \cos 40°)$$

$$= \sin^2 20° + \frac{1}{4} + \frac{1}{2}\cos 40°$$

$$= \frac{1}{2}(1 - \cos 40°) + \frac{1}{4} + \frac{1}{2}\cos 40°$$

$$= \frac{1}{2} + \frac{1}{4}$$

$$= 0.75.$$

Answer: 0.75

Problem 10 Solution

Let $y = \log_{10} x$, then

$$\log_{\sqrt{10}} x = \frac{\log_{10} x}{\log_{10} \sqrt{10}} = 2y.$$

So the equation becomes

$$2|y - 1| - |y - 2| = 2.$$

Solve this equation in 3 cases:

Case (1): $y \leq 1$. Then

$$-2(y - 1) + (y - 2) = 2,$$

so $y = -2$.

Case (2): $1 < y \leq 2$. Then

$$2(y-1) + (y-2) = 2,$$

thus $y = 2$.

Case (3): $y > 2$. Then

$$2(y-1) - (y-2) = 2,$$

we get $y = 2$, but this solution is out of the range $y > 2$.

Hence there are two solutions: $y = -2$ and $y = 2$. So the solutions for x are: $x = 0.01$ and $x = 100$, thus the difference is

$$100 - 0.01 = 99.99.$$

Answer: 99.99

Problem 11 Solution

For any integer a, $a^2 \equiv 0,1,2,4 \pmod 7$, and after some simple checks, we find that $7 \mid (a^2 + b^2)$ only if $a \equiv b \equiv 0 \pmod 7$. There are 14 multiples of 7 between 1 and 100. To select the pair (a,b), there are two cases: (1) $a = b$, there are 14 choices; (2) $a < b$, there are $\binom{14}{2} = 91$ choices.

Thus the final answer is $14 + 91 = 105$.

Answer: 105

Problem 12 Solution

After 5 jumps, the frog must be at B, D, or F (this can be proven by coloring: color the vertices A, C, E with red, and the vertices B,

D, F with black. At each jump he alternates between the colors, so after an even number of jumps he ends up at a red vertex and after an odd number of jumps he ends up at a black vertex).

If the frog reaches D in fewer than 5 jumps, the only possible way is that he makes 3 jumps along one direction, clockwise or counterclockwise, to get to D. There are 2 ways in this case.

Now suppose the frog does not reach D in 3 jumps. So there are 2 choices at each jump, and there are $2^3 - 2 = 6$ ways for the first 3 jumps so that he does not reach D. Then he can jump randomly for the remaining 2 jumps, $2^2 = 4$ ways. In total, $6 \times 4 = 24$ ways in this case.

Adding the two cases together, there are $2 + 24 = 26$ ways in total.

Answer: 26

Problem 13 Solution
Let
$$A = \frac{c}{a} + \frac{a}{b+c} + \frac{b}{c},$$
then by AM-GM Inequality,

$$A + 1 = \frac{c}{a} + \frac{a}{b+c} + \frac{b+c}{c} \geq 3 \cdot \sqrt[3]{\frac{c}{a} \cdot \frac{a}{b+c} \cdot \frac{b+c}{c}} = 3,$$

thus
$$A \geq 2.$$

Equality occurs when
$$\frac{c}{a} = \frac{a}{b+c} = \frac{b+c}{c},$$
which means $a = c > 0$ and $b = 0$.

Answer: 2

Problem 14 Solution

Suppose N can be written as the sum of two or more consecutive positive integers as follows.

$$
\begin{aligned}
N &= m + (m+1) + (m+2) + \cdots + (m+k) \\
&= (k+1)m + \frac{k(k+1)}{2} \\
&= \frac{1}{2}(k+1)(2m+k).
\end{aligned}
$$

The two factors $k+1$ and $2m+k$ have opposite parity: one is even and the other is odd, and both are greater than 1. So we conclude that the number N has at least one odd factor greater than 1.

On the other hand, suppose N has at least one odd factor greater than 1. Let this odd factor be $2k+1$, and $N = (2k+1)m$ where k and m are positive integers. We now verify that N can be written as the sum of two or more consecutive positive integers. If $mr > k$, then

$$
N = (m-k) + (m-k+1) + \cdots + m + (m+1) + \cdots + (m+k).
$$

If $m \le k$, then

$$
N = (k-m+1) + (k-m+2) + \cdots + (k+m-1) + (k+m).
$$

Therefore, N can be written as the sum of two or more consecutive positive integers if and only if N has at least one odd factor greater than 1.

Consequently, if N cannot be written as the sum of two or more consecutive positive integers, it can only has 2 as its prime factor. Thus, $N = 2^p$ where $p \ge 0$ is an integer. There are 10 powers of 2 under 1000: 2^0, 2^1, 2^2, ..., 2^9. Therefore the answer is 10.

Answer: 10

Problem 15 Solution

For the two curves to have at least one common point, the equation

$$2y = 4 - 4\left(y - \frac{a}{2}\right)^2$$

should have at least one nonnegative root for y. Simplify the equation,

$$2y + 4y^2 - 4ay + a^2 = 4,$$

so

$$4y^2 + (2 - 4a)y + (a^2 - 4) = 0.$$

First the discriminant

$$(2 - 4a)^2 - 4 \cdot 4 \cdot (a^2 - 4) \geq 0,$$

which is

$$4 - 16a + 16a^2 - 16a^2 + 64 \geq 0,$$

so

$$a \leq \frac{17}{4} = 4.25.$$

To find the range of a for which the equation has at least one nonnegative root, we first fine the opposite scenario: both roots are negative. By Vieta's theorem, this means

$$a^2 - 4 > 0 \text{ and } 2 - 4a > 0,$$

and the solution for these inequalities is

$$a \in ((-\infty, -2) \cup (2, \infty)) \cap (-\infty, 1/2)$$

which simplifies to $a < -2$. So if there should be at least one nonnegative root

$$-2 \leq a \leq 4.25.$$

Thus $M = 4.25$ and $N = -2$, and $M - N = 6.25$.

Answer: 6.25

Problem 16 Solution

Since adjacent faces are painted with different colors, at least 3 colors are needed.

Case (1): All 6 colors are used. Fix one color to be the top face. There are 5 choices for the bottom face. The 4 colors for the 4 side faces form a circular permutation, so there are $(4-1)! = 6$ ways. In total, there are 30 ways for this case.

Case (2): 5 colors are used. One of the colors is repeated: 6 choices for that; the two faces with that color are opposite faces. Let those 2 faces be the top and bottom faces. Choose 4 colors for the remaining faces: $\binom{5}{4} = 5$ ways. The remaining 4 colors form a circular permutation, and reversing order is equivalent to flipping the cube upside-down, so there are $6 \times 5 \times (4-1)! \div 2 = 90$ ways for this case.

Case (3): 4 colors are used. Two of the colors are repeated: $\binom{6}{2} = 15$ choices. Since faces with the same color have to be opposite faces, there is only one way to place the two pairs of faces with the selected colors, up to rotation. The remaining 2 faces use 2 of the remaining 4 colors, and there is no difference as to which remaining faces to have which color, so the number of choices is $\binom{4}{2} = 6$. Thus the number of ways for this case is $15 \times 6 = 90$.

Case (4): 3 colors are used. To select the 3 colors, there are $\binom{6}{3} = 20$ ways. Once the 3 colors are chosen, there is only one way to paint the faces since any painting methods can be rotated to become identical. So there are 20 ways for this case.

In total, there are $30 + 90 + 90 + 20 = 230$ ways.

Answer: 230

Problem 17 Solution

The sequence appears to satisfy a second-order recurrence relation, with $3 \pm 2\sqrt{2}$ as the roots of the characteristic equation. Also

$$(3 + 2\sqrt{2}) + (3 - 2\sqrt{2}) = 6, \quad (3 + 2\sqrt{2})(3 - 2\sqrt{2}) = 1,$$

so the characteristic equation is

$$x^2 - 6x + 1 = 0.$$

Thus the recurrence relation is

$$a_{n+2} = 6a_{n+1} - a_n, \quad n \geq 0.$$

Note: it is also easy to verify directly that $a_{n+2} - 6a_{n+1} = a_n$.

For $n = 0$, $a_0 = 1$; for $n = 1$, $a_1 = 3$. Based on the recurrence relation, we calculate the last digit of each term, for $n = 0, 1, 2, 3, \ldots$:

$$1, 3, 7, 9, 7, 3, 1, 3, 7, 9, 7, 3, \ldots$$

and it cycles with period 6. Since $2018 \equiv 2 \pmod 6$, the answer is 7.

Answer: 7

Problem 18 Solution

Let

$$n = \overline{d_1 d_2 \ldots d_{k-1} d_k} = m^2$$

be a k digit number, thus $d_1 d_k \neq 0$. Also,

$$\overline{d_1 d_2 \ldots d_{k-2}} = p^2$$

where p is a positive integer. Since

$$(10p)^2 = \overline{d_1 d_2 \ldots d_{k-2} 00} < m^2,$$

we get $m > 10p$, which means $m \geq 10p + 1$. We also have

$$m^2 - (10p)^2 = \overline{d_{k-1} d_k} \leq 99,$$

thus

$$(10p + 1)^2 \leq m^2 \leq (10p)^2 + 99,$$

so

$$100p^2 + 20p + 1 \leq 100p^2 + 99,$$

which simplifies to

$$10p \leq 49,$$

hence, $p \leq 4$. Since we are looking for the largest n, let $p = 4$, so

$$41^2 \leq m^2 \leq 40^2 + 99,$$

then

$$1681 \leq m^2 \leq 1699.$$

There is exactly one perfect square in this range: 1681, so the largest value of n is 1681.

Answer: 1681

Problem 19 Solution

Let $a = PA$, $b = PB$, and $c = PC$. Then

$$AB = \sqrt{a^2 + b^2}, \quad BC = \sqrt{b^2 + c^2}, \quad CA = \sqrt{c^2 + a^2},$$

and the sum of edge lengths

$$S = a + b + c + \sqrt{a^2 + b^2} + \sqrt{b^2 + c^2} + \sqrt{c^2 + a^2} = 18,$$

and also the volume

$$V = \frac{abc}{6}.$$

Using the AM-GM Inequality multiple times,

$$a+b+c \geq 3\sqrt[3]{abc},$$

and

$$
\begin{aligned}
& \sqrt{a^2+b^2} + \sqrt{b^2+c^2} + \sqrt{c^2+a^2} \\
\geq{}& \sqrt{2ab} + \sqrt{2bc} + \sqrt{2ca} \\
\geq{}& 3\sqrt[3]{\sqrt{2ab} \cdot \sqrt{2bc} \cdot \sqrt{2ca}} \\
={}& 3\sqrt{2}\sqrt[3]{abc}.
\end{aligned}
$$

Equality occurs when $a = b = c$ in each of the inequalities above. Thus

$$
\begin{aligned}
S &\geq 3\sqrt[3]{abc} + 3\sqrt{2}\sqrt[3]{abc} \\
&= 3(1+\sqrt{2})\sqrt[3]{abc} \\
&= 3(1+\sqrt{2})\sqrt[3]{6V},
\end{aligned}
$$

and equality occurs when $a = b = c$. Therefore

$$
\begin{aligned}
V &\leq \frac{1}{6}\left(\frac{S}{3(1+\sqrt{2})}\right)^3 \\
&= \frac{1}{6} \cdot 6^3(\sqrt{2}-1)^3 \\
&= 36(5\sqrt{2}-7) \\
&= 180\sqrt{2} - 252.
\end{aligned}
$$

Therefore $K = 180$, $M = 2$, and $N = 252$, and

$$K+M+N = 434.$$

Answer: 434

Problem 20 Solution

There are 900 three-digit numbers. We count the number of cards that can be turned to represent different numbers. So the

---

first and last digits can be $1, 6, 8, 9$, and the middle digits can be $0, 1, 6, 8, 9$. Therefore the total number of cards that can be turned is $4 \times 5 \times 4 = 80$.

However, among these numbers, some just turn into themselves. These include the following cases: the first anf last digits are 1 and 1, 8 and 8, 6 and 9, 9 and 6. The middle digit can be 0, 1, or 8. So there are $4 \times 3 = 12$ such numbers.

Thus, the number of cards that can be saved is $(80 - 12) \div 2 = 34$. So the number of cards that must be printed is $900 - 34 = 866$.

Answer: 866

2.9 ZIML June 2018 Varsity Division

Below are the solutions from the Varsity Division ZIML Competition held in June 2018.
The problems from the contest are available on p.65.

Problem 1 Solution
The equation can be written as

$$\sqrt{(x-4)^2+3^2}+\sqrt{(x-5)^2+2^2}=\sqrt{26}.$$

The left hand side is the sum of distances from the point $(x,0)$ to the points $(4,3)$ and $(5,-2)$. By the triangle inequality,

$$\sqrt{(x-4)^2+3^2}+\sqrt{(x-5)^2+2^2} \geq \sqrt{(4-5)^2+(3+2)^2}$$
$$= \sqrt{26},$$

thus equality holds, so the point $(x,0)$ must be on the line segment between $(4,3)$ and $(5,-2)$. The line equation between these two points is (by two-point formula)

$$\frac{x-4}{5-4}=\frac{y-3}{-2-3},$$

which simplifies to

$$y=-5x+23.$$

Let $y=0$, we get $x=\dfrac{23}{5}=4.6$. This is the only solution.

Answer: 4.6

Problem 2 Solution
The point M is the midpoint of $\overline{A_1C_1}$, and N is the midpoint of $\overline{B_1C}$, and the two planes DA_1C_1 and ACB_1 are parallel, so the distance between the two skew lines $\overline{AN}$ and $\overline{DM}$ is equal to the distance between the two planes mentioned above. Also, both planes are perpendicular to the space diagonal $\overline{BD_1}$.

Let P be the center of equilateral triangle ACB_1, and Q be the center of equilateral triangle DA_1C_1, then P and Q are both on the space diagonal $\overline{BD_1}$, and PQ is the desired distance between the two skew lines $\overline{AN}$ and $\overline{DM}$.

Since $AB = \sqrt{3}$, the space diagonal $BD_1 = \sqrt{3} \cdot \sqrt{3} = 3$. To find PQ, we first find BP and D_1Q (which are equal by symmetry). $\overline{BP}$ is the altitude of tetrahedron B-ACB_1 on the base ACB_1, so we use the volume method to find it. The volume of tetrahedron B-ACB_1 is
$$(\sqrt{3})^3/6 = \sqrt{3}/2.$$
Also $AC = \sqrt{2}AB = \sqrt{6}$, thus
$$\frac{1}{3} \cdot \frac{\sqrt{3}}{4}(\sqrt{6})^2 \cdot BP = \frac{\sqrt{3}}{2},$$
simplifying and solving for BP,
$$BP = 1.$$
Therefore $D_1Q = 1$, and
$$PQ = BD_1 - BP - D_1Q = 3 - 1 - 1 = 1.$$

Answer: 1

Problem 3 Solution

Let $\oplus$ denote testing positive. We want $P(D|\oplus) \geq 90\%$. If we let $p = P(\oplus|D)$ (so the false negative rate is $1 - p$) we have

$$
\begin{aligned}
P(D|\oplus) &= \frac{P(D \cap \oplus)}{P(\oplus)} \\
&= \frac{P(D) \cdot P(\oplus|D)}{P(D) \cdot P(\oplus|D) + P(D^c) \cdot P(\oplus|D^c)} \\
&= \frac{40\% \cdot p}{40\% \cdot p + 60\% \cdot 5\%}.
\end{aligned}
$$

Solving $0.90 = \dfrac{0.4p}{0.4p+0.03}$ we get $0.36p + 0.027 = 0.4p$ and

hence $p = \dfrac{27}{40} = 0.675$. Hence as long as the false negative rate is less than or equal to 32.5% the goal will be achieved.

Answer: 32.5

Problem 4 Solution

Let $x = \overline{ab}$ and $y = \overline{cd}$. Then we are looking for integer solutions of $(x+y)^2 = 100x + y$, where $10 \le x \le 99$ and $0 \le y \le 99$. Thinking of the equation as a quadratic equation on x it becomes

$$x^2 + (2y - 100)x + (y^2 - y) = 0,$$

which has discriminant

$$\Delta = (2y - 100)^2 - 4(y^2 - y) = 10000 - 396y = 4(2500 - 99y).$$

We need this discriminant to be a positive perfect square, thus $2500 - 99y = k^2$ for some positive integer k. Factoring, we have

$$3^2 \cdot 11 \cdot y = (50 + k)(50 - k).$$

Note it is not possible for both $50 + k$ and $50 - k$ to be multiples of 3, as this would imply 100 is a multiple of 3. Hence we have 3 cases: (i) $9|50 - k$ and $11|50 + k$; (ii) $11|50 - k$ and $9|50 + k$; or (iii) $99|50 + k$.

For (i), if

$$50 - k = 0, 9, 18, 27, 36, 45,$$

then

$$50 + k = 100, 91, 82, 73, 64, 55,$$

respectively, thus $50 - k = 45$, $50 + k = 55$ and $k = 5$; therefore $y = 25$ and $x = 20, 30$. Proceeding similarly for (ii) we see there are no solutions for y. For (iii) we have $50 + k = 99$, thus $y = 1$

---

and $x = 98$. Therefore the only solutions are 2025, 3025, and 9801, which add up to 14851.

Answer: 14851

Problem 5 Solution

We know that $a = h + d$, $b = h + 2d$, and $c = h + 3d$. Then the semiperimeter

$$s = \frac{a + b + c}{2} = \frac{3h + 6d}{2}.$$

Based on Heron's formula, the area of the triangle is

$$\sqrt{s(s-a)(s-b)(s-c)} = \sqrt{\frac{3h + 6d}{2} \cdot \frac{h + 4d}{2} \cdot \frac{h + 2d}{2} \cdot \frac{h}{2}},$$

and the area of the triangle can also be calculated as

$$\frac{bh}{2} = \frac{(h + 2d)h}{2},$$

thus

$$\frac{3h + 6d}{2} \cdot \frac{h + 4d}{2} \cdot \frac{h + 2d}{2} \cdot \frac{h}{2} = \frac{(h + 2d)^2 h^2}{4},$$

$$\frac{3(h + 2d)^2(h + 4d)h}{16} = \frac{(h + 2d)^2 h^2}{4},$$

$$3(h + 4d) = 4h,$$

$$h = 12d.$$

Therefore, $\dfrac{h}{d} = 12$.

Answer: 12

Problem 6 Solution

Let the middle number $a_{51} = x$, then $a_1 = x - 50$, $a_2 = x - 49$, ..., and the sequence is

$$x - 50, x - 49, \ldots, x - 1, x, x + 1, \ldots, x + 49, x + 50.$$

So the equation becomes

$$(x-50)^2 + (x-49)^2 + \cdots + (x-1)^2 + x^2$$
$$= (x+1)^2 + (x+2)^2 + \cdots + (x+50)^2,$$

which is

$$51x^2 - 2x \sum_{k=1}^{50} k + \sum_{k=1}^{50} k^2 = 50x^2 + 2x \sum_{k=1}^{50} k + \sum_{k=1}^{50} k^2,$$

and this simplifies to

$$x^2 - 5100x = 0.$$

Since $x \neq 0$, we get $x = 5100$, so $a_1 = x - 50 = 5050$. This is the only possible value.

Answer: 5050

Problem 7 Solution
We use complementary counting. In total there are $2^{7-1} = 64$ sequences with a sum of 7. (Consider $1+1+1+1+1+1+1$ with 6 plus signs, each can either stay or be removed.)

We then count sequences that are not pyramidal. For such a sequence to not be pyramidal, it must have at least 3 terms with 2 of those terms greater than one. The only such sums for 7 are

$$7 = 4+2+1 = 3+3+1 = 3+2+2$$
$$= 3+2+1+1 = 2+2+2+1 = 2+2+1+1+1.$$

We consider each as separate cases.

For $4+2+1$ we have $(4,1,2)$ or its reverse. For $3+1+3$ we only have $(3,1,3)$. For $3+2+2$ no rearrangement works. For $3+2+1+1$ we have $(3,1,2,1),(3,1,1,2),(1,3,1,2)$ or their reverses. For $2+2+2+1$ we have $(2,2,1,2)$ or its reverse.

Lastly for $2 + 2 + 1 + 1 + 1$ we have $(2,1,2,1,1), (2,1,1,2,1)$ and their reverses as well as $(2,1,1,1,2)$ and $(1,2,1,2,1)$.

Hence there are $64 - 2 - 1 - 0 - 6 - 2 - 6 = 47$ total such pyramidal sequences. (Note, for the purposes of the problem pyramidal was used, but these sequences are often called unimodal sequences.)

Answer: 47

Problem 8 Solution
Writing 2101_b in place values,

$$2101_b = 2b^3 + b^2 + 1 = (2b^2 - b + 1)(b + 1).$$

Since $2b^2 - b + 1 = (2b - 3)(b + 1) + 4$, the greatest common divisor of $2b^2 - b + 1$ and $b + 1$ is a factor of 4, so it can be 1, 2, or 4.

If $\gcd(2b^2 - b + 1, b + 1) = 1$ or 4, and since their product is a perfect square, both have to be a perfect square. Thus

$$b + 1 = n^2, \quad n \in \mathbb{N}, 2 \leq n \leq 10.$$

For $n = 2$, $b = 3$, and $2b^2 - b + 1 = 16$ is a perfect square, so $b = 3$ is a solution;

For $n = 3$, $b = 8$, and $2b^2 - b + 1 = 121$ is a perfect square, so $b = 8$ is a solution;

Checking $n = 4, 5, \ldots, 10$, none of them satisfies the requirement.

If $\gcd(2b^2 - b + 1, b + 1) = 2$, each of $2b^2 - b + 1$ and $b + 1$ is 2 times a perfect square. So

$$b + 1 = 2n^2, \quad n \in \mathbb{N}, 2 \leq n \leq 7.$$

Checking these values for n, none of them produces solutions.

Therefore the only possible bases are $b = 3$ and $b = 8$, hence the final answer is $3 + 8 = 11$.

Answer: 11

Problem 9 Solution

Preserving horizontal and vertical symmetries it is enough to determine the five colors A, B, C, D, E shown below.

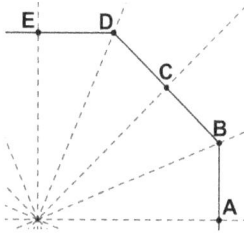

Label the three symmetries we want to break $22.5°, 45°, 67.5°$. To break the $22.5°$ symmetry we must have $A \neq C$. Similarly to break the $67.5°$ symmetry, we must have $C \neq E$. To break the $45°$ symmetry, $B \neq D$ or $B = D$ and $A \neq E$. Consider these as two separate cases.

If $B \neq D$, we have $4 \cdot 3 = 12$ choices for B and D. Then for A, C, E choosing C first (and then A and E) we have $4 \cdot 3 \cdot 3 = 36$ choices. Hence in this case there are $12 \cdot 36 = 432$ outcomes.

If $B = D$ and $A \neq E$, we have 4 choices for B (and D). Then for A, C, E there are $4 \cdot 3 \cdot 2 = 24$ choices. Hence in this case there are $4 \cdot 24 = 96$ outcomes.

Therefore we have $432 + 96 = 528$ total outcomes.

Answer: 528

Problem 10 Solution

Let x be an integer root of the given equation. Then

$$a = \frac{2 - 3x}{x^2 - x + 1}.$$

Clearly $a = 0$ is not a possible value.

Also, $x^2 - x + 1 = (x - 1/2)^2 + 3/4 > 0$ for all $x \in \mathbb{R}$.

If $a > 0$, which means $a \geq 1$,

$$2 - 3x \geq x^2 - x + 1,$$

so

$$-1 - \sqrt{2} \leq x \leq -1 + \sqrt{2},$$

hence the possible values are $x = -2, -1, 0$, and only $x = 0$ gives an integer $a = 2$.

If $a < 0$, which means $a \leq -1$,

$$2 - 3x \leq -x^2 + x - 1,$$

so

$$1 \leq x \leq 3,$$

hence the possible values are $x = 1, 2, 3$, and when $x = 1$ or 3 we get the same integer value $a = -1$.

Therefore, the sum of all possible values of a is $2 + (-1) = 1$.

Answer: 1

Problem 11 Solution

It is impossible for any of the two numbers to be equal (otherwise, if $a = b$, then $a \mid (cb + 1)$ implies $a = 1$, not possible).

Without loss of generality, assume $a < b < c$. Since $b \mid (ca + 1)$ and $c \mid (ab + 1)$, we get

$$bc \mid (ca + 1)(ab + 1),$$

that is

$$bc \mid a^2bc + ab + ac + 1,$$

hence

$$bc \mid ab + ac + 1.$$

Let $ab + ac + 1 = kbc$ where k is a positive integer. Since $ab < bc$, we have $ab + 1 \le bc$ because both sides are integers. Also $ac < bc$, thus

$$ab + ac + 1 < 2bc,$$

then

$$kbc \le 2bc,$$

therefore $k = 1$. Hence

$$ab + ac + 1 = bc.$$

Given that $a \mid (bc + 1)$, we have

$$a \mid (ab + ac + 2),$$

thus $a \mid 2$. But $a > 1$, so $a = 2$.

Then, $2b + 2c + 1 = bc$.

From $c \mid (ab + 1)$ we get $c \mid (2b + 1)$. Also, from $b < c$, we get $2b + 1 < 2c$, therefore $c = 2b + 1$.

So

$$2b + 2(2b + 1) + 1 = b(2b + 1),$$
$$0 = 2b^2 - 5b - 3,$$

so $b = 3$ (throwing away negative root), and then $c = 7$.

Therefore, there is only one possible value for abc, which is $2 \times 3 \times 7 = 42$.

Answer: 42

Problem 12 Solution

Since C' is the reflection of C about $\overline{BD}$,

$$\angle C'BD = \angle CBD = \angle ADB,$$

thus $BE = DE$. Let $DE = x$, then $BE = x$ and

$$AE = AD - DE = 8 - x,$$

so

$$x^2 = 4^2 + (8-x)^2,$$

solve and get $x = 5$. Therefore the area of $\triangle BED$ equals

$$\frac{1}{2} \cdot DE \cdot AB = \frac{1}{2} \cdot 5 \cdot 4 = 10.$$

Answer: 10

Problem 13 Solution

Note that $\sqrt[3]{x^2-4} = \sqrt[3]{x+2} \cdot \sqrt[3]{x-2}$, so the equation can be rearranged and factored:

$$\sqrt[3]{x^2-4} - 2\sqrt[3]{x+2} - 2\sqrt[3]{x-2} + 4 = 0,$$

$$(\sqrt[3]{x+2} - 2)(\sqrt[3]{x-2} - 2) = 0.$$

If $\sqrt[3]{x+2} - 2 = 0$, we get $x = 6$; if $\sqrt[3]{x-2} - 2 = 0$, we get $x = 10$. Both are solutions after checking. So the sum of all real roots is $6 + 10 = 16$.

Answer: 16

Problem 14 Solution

Let E be the intersection of $\overline{AC}$ and $\overline{BD}$. Let $a = AB$, $x = AE$, and $y = DE$, so

$$\frac{2x}{a} = \frac{a}{2y},$$

which is

$$a^2 = 4xy.$$

Also we have

$$a^2 = x^2 + y^2,$$

thus

$$x^2 + y^2 = 4xy.$$

Given that $AC > BD$, we have $x > y$, then

$$\frac{y}{x} = 2 - \sqrt{3}.$$

This means that $\tan \angle DAE = 2 - \sqrt{3}$. Using double-angle formula,

$$\tan \angle DAB = \tan 2\angle DAE = \frac{2 \tan \angle DAE}{1 - \tan^2 \angle DAE} = \frac{4 - 2\sqrt{3}}{4\sqrt{3} - 6} = \frac{1}{\sqrt{3}},$$

therefore

$$\angle DAB = 30°.$$

Answer: 30

Problem 15 Solution

First note that at most one queen can be in a corner (as the other 3 corners are in the same row, column, or diagonal). Therefore first assume a queen is in the top left corner. We can then restrict our attention to a 3×3 square. For two queens to be placed in this square, they must in one of the four arrangements shown below:

Q			
		Q	
	Q		

Q			
		Q	
			Q

Q			
		Q	
	Q		

Q			
			Q
	Q		

With rotations this gives a total of $4 \times 4 = 16$ arrangements. Further, avoiding the corners, note it is impossible to place a queen in one of the four center squares:

x			x
	Q		
			$?$
x		$?$	x

as the only remaining possibilities are in the same diagonal. Hence we are left with arrangements avoiding the corners and avoiding the four center squares. Any of these arrangements, however, can be created by removing one of the four queens in the two examples given in the problem. Therefore we have $2 \times 4 = 8$ additional arrangements, giving a final answer of

$$16 + 8 = 24$$

Answer: 24

Problem 16 Solution

Define a sequence $\{b_k\}$ as follows:

$$b_k = \begin{cases} 0, & \text{if } a_k \text{ is even;} \\ 1, & \text{if } a_k \text{ is odd.} \end{cases}$$

Then the sequence $\{b_k\}$ is

$$0, 0, 1, 0, 0, 0, 1, 1, 1, 1, 0, 1, 0, 1, 1, 0, 0, 1, 0, 0, 0, 1, 1, 1, 1, \ldots$$

It is clear to see the pattern $b_{i+15} = b_i$ for all $i \in \mathbb{N}$.

Since $2018 \equiv 8 \pmod{15}$, $b_{2018} = b_8 = 1$, and so $b_{2019} = 1$ and $b_{2020} = 1$, therefore a_{2018}, a_{2019}, and a_{2020} are all odd numbers. Hence

$$a_{2018}^2 + a_{2019}^2 + a_{2020}^2 \equiv 1 + 1 + 1 \equiv 3 \pmod{4}.$$

Answer: 3

Problem 17 Solution

Note we have 6 groups of white and blue in total, so the red cards must be separated by these groups. That is the row of cards is

$$R(-)R(-)R(-)R(-)R(-)R(-)R$$

where each $(-)$ denotes one or more white cards or one or more blue cards. There are $\binom{6}{3} = 20$ ways to decide which of the 6 groups are white and which are blue. Then for each of these groups we use the positive version of stars and bars to count the number of ways to place the 7 cards into the 3 spots (with each spot having at least one card). This gives

$$\binom{7-1}{3-1} \cdot \binom{7-1}{3-1} = 15^2 = 225.$$

Hence the number of total arrangements is $20 \cdot 225 = 4500$.

Answer: 4500

Problem 18 Solution

Use the identity $\sin^2 \theta = 1 - \cos^2 \theta$ and apply the AM-GM Inequality,

$$
\begin{aligned}
f(\theta) &= (1+\cos\theta)^2(1-\cos^2\theta) \\[2mm]
&= (1+\cos\theta)^3(1-\cos\theta) \\[2mm]
&= \frac{1}{3}(1+\cos\theta)^3(3-3\cos\theta) \\[2mm]
&\le \frac{1}{3}\left(\frac{3(1+\cos\theta)+(3-3\cos\theta)}{4}\right)^4 \\[2mm]
&= \frac{1}{3}\cdot\left(\frac{3}{2}\right)^4 \\[2mm]
&= \frac{27}{16},
\end{aligned}
$$

where equality occurs when $1+\cos\theta = 3 - 3\cos\theta$, which means $\cos\theta = \dfrac{1}{2}$, that is $\theta = 2k\pi \pm \dfrac{\pi}{3}$.

Therefore the final answer is $\dfrac{27}{16} = 1.6875 \approx 1.69$.

Answer: 1.69

Problem 19 Solution

It is easy to see that if (x,y) is a solution, so is $(-x,-y)$. Thus we only focus on $y \ge 0$.

The equation can be treated as a quadratic equation in x:
$$x^2 + yx + (4y^2 - 61) = 0.$$

For this equation to have integer solutions, the discriminant should be a perfect square:
$$y^2 - 4(4y^2 - 61) = k^2, \quad k \ge 0 \text{ is an integer.}$$

that is
$$15y^2 + k^2 = 244,$$

thus
$$y^2 \le \sqrt{244/15}$$

and since y is an integer,

$$y^2 \le 16.$$

So we try $y = 0, 1, 2, 3, 4$, and only $y = 4$ gives integer solutions for x:
$$x^2 + 4x + 64 = 61$$

and $x = -1$ or $x = -3$.

Hence the solutions are $(-1,4)$, $(-3,4)$, $(1,-4)$, and $(-3,4)$, and the largest value for $|x| + |y|$ is $3 + 4 = 7$.

Answer: 7

Problem 20 Solution
If $x \le \dfrac{y}{x^2 + y^2}$, then $z = x$, and

$$z^2 \le \frac{xy}{x^2 + y^2} \le \frac{xy}{2xy} = \frac{1}{2}.$$

If $\dfrac{y}{x^2 + y^2} \le x$, then $z = \dfrac{y}{x^2 + y^2}$, so

$$z^2 \le \frac{xy}{x^2 + y^2} \le \frac{xy}{2xy} = \frac{1}{2}.$$

In any case, the maximum value is $\dfrac{1}{2} = 0.5$.

Answer: 0.5

3. Appendix

3.1 Varsity Division Topics Covered

Algebra

- Students should be comfortable with ratios, proportions, and their applications to problems involving work and motion, but these problems are not a main focus at this level
- Radicals, Exponents, and Logarithms: Simplest Radical Form for Roots, Laws of Exponents, Laws of Logarithms including change of base
- Complex Numbers: Arithmetic Operations, Rectangular, Polar, and Exponential Forms, De Moivre's identity, Roots of Unity
- Factoring Polynomials: Sums and differences of squares, cubes, etc., Binomial and multinomial theorem, Completing the Square/Rectangle, etc.
- Solving Equations and Inequalities: Linear Equations and Inequalities, Quadratic Equations and Inequalities, Systems of Equations, Change of Variables, Equations and Inequalities involving Radicals, Absolute Values, Fractions, etc.

- Quadratics: Graphing and Vertex Form, Maxima and Minima, Quadratic Formula, Discriminant, Vieta's Theorem for sum and product of the roots
- Polynomials: Polynomial Long Division, Remainder and Factor Theorem, Rational Root Theorem, General Vieta's Theorem, Fundamental Theorem of Algebra
- Inequalities: finding maximum and minimum of algebraic expressions, Arithmetic-Mean-Geometric-Mean (AM-GM) Inequality, Cauchy-Schwarz Inequality, Rearrangement Inequality, Weighted AM-GM, Jensen's Inequality, Power Means, Hölder's Inequality, etc.
- Functional Equations: Cauchy type equations, finding function values, fixed points, periods, bounds and other properties based on given functional equations

Geometry

- As a general rule students should be comfortable using algebraic techniques (linear equations, quadratic equations, systems of equations, etc.) as tools for applying the geometric concepts listed below
- Angles in Parallel Lines (corresponding angles, alternating interior/exterior angles, same-side interior/exterior angles, etc.)
- Analytical Geometry: Equations of Lines, Parabolas, and Circles, Distance Formula, Midpoint Formula, Geometric Interpretation of Slope and Angles
- Triangles: Congruence and Similarity, Pythagorean theorem, Ratios of Sides for triangles with angles of $45°, 45°, 90°$ or $30°, 60°, 90°$
- Trigonometry: General understanding of sine, cosine, tangent, and their co-functions, Laws of Sines and Cosines, Trigonometric Identities

- Centers in Triangles: Definitions of altitudes, medians, angle bisectors, perpendicular bisectors, Definitions and basic properties of orthocenter, centroid, incenter, circumcenter, Angle Bisector Theorem, Concurrence and Collinearity, Ceva's and Menelaus's Theorems
- Interior and Exterior Angles of Polygons, including the sum of all these angles, each angle if the polygon is regular, etc.
- Areas and Perimeters of basic shapes such as triangles, rectangles, parallelograms, trapezoids, and circles, Heron's formula and formulas using inradius or circumradius for triangles
- Geometric Reasoning with Areas: Congruent shapes have the same area, Similar triangles have a ratio of areas that is the square of the ratio of their sides, Triangles with the same height have a ratio of their areas equal to the ratio of their bases, etc., Using multiple expressions of area to solve for unknowns
- Circles: Arc Length, Sector Area, Definitions for Tangent Lines and Tangent Circles, Inscribed Angles, Angles formed by intersecting chords, Power of a Point, Ptolemy's Theorem
- Lines and Plane in 3-D: Definitions of parallel, intersecting, and skew lines, Definitions of parallel and intersecting planes, Calculation of angles between lines and/or planes in 3-D space
- Solid Geometry: Surface Area and Volume for Spheres, Prisms, Pyramids, and Cones, Reasoning for more general solids, such as combining the solids listed above or pieces of solids when cut by a plane, etc.; Dihedral Angles, Polyhedral Angles, Platonic Solids, Euler's Polyhedral Formula

Counting and Probability

- Fundamental Rules: Sum and Product Rules, Permutations and Combinations

- Counting Methods: Complementary counting, Stars and bars (also called sticks and stones, balls and urns, etc.), Grouping objects that must be together, Inserting objects that must be apart into spaces between objects, etc., Principle of Inclusion and Exclusion
- Identities: Symmetry, Pascal's Identity, Hockey Stick Identity, etc. for binomial coefficients, Binomial and Multinomial Theorems, Understanding of these identities using combinatorial proofs
- Sequences: Arithmetic and Geometric Sequences and Series, Finding and understanding patterns and recursive definitions for general sequences
- Probability and Sets: Definitions for event, sample space, complement, intersection, and union, Understanding the use of Venn Diagrams
- Probability: In finite sample spaces as a ratio of the number of outcomes, In geometric sample spaces as a ratio of lengths, areas, or volumes, Axioms of Probability, Independence, Conditional Probability, Law of Total Probability and Bayes's Theorem
- Probability Distributions: Definitions and Understandings of Probability Distributions and Expected Value

Number Theory

- Fundamental Definitions: Prime numbers, factors/divisors, multiples, least common multiple (LCM), greatest common factor/divisor (GCF or GCD), perfect squares/cubes/etc.
- Number Bases: Expressing and converting numbers in base 2, 3, 8, 16, etc, Understanding how to perform arithmetic in different bases
- Divisibility Rules for numbers such as 2, 3, 4, 5, 8, 9, 10, 11, and how to combine the rules for numbers such as 6, 22, etc.

- (Unique) Prime Factorization and how to use the prime factorization to find the number of factors, to test whether a number is a perfect square/cube/etc, to find the LCM or GCD.
- Factoring Tricks: Factors come in pairs, perfect squares have an odd number of factors, etc.
- Modular Arithmetic: Connection with remainders and applications such as "find the units digit", General rules for addition, subtraction, multiplication, and division, Extension of divisibility rules to calculating a number modulo 9, 11, etc., Fermat's Little Theorem, Euler's Totient Function and extension to Fermat's Little Theorem, Wilson's Theorem, Chinese Remainder Theorem
- Gauss' Floor function $\lfloor x \rfloor$: properties, equation and inequalities involving the Floor function, Fractional part function $\{x\}$, solving problems using these functions
- Modular equations: general modular equations, Quadratic Residue and non-residue, Euler's Criterion, Legendre symbol, Quadratic Reciprocity

3.2 Glossary of Common Math Terms

Acute Angle An angle less than $90°$.

Altitude of a Triangle A line segment connecting a vertex of a triangle to the opposite side forming a right angle. Also called the height of a triangle.

Angle A figure formed by two rays sharing a common vertex. Often measured in degrees.

Angle Bisector A line dividing an angle into two equal halves.

Arc The curve of a circle connecting two points.

Area The amount of space a region takes up. Often denoted using square brackets: area of $\triangle ABC = [ABC]$.

Arithmetic Sequence A sequence where the difference between one term and the next is constant.

Average See Mean.

Base of a Triangle One side of a triangle, often used when the altitude is drawn from the opposite side to this base.

Binomial Coefficient The symbol $\binom{n}{k} = \dfrac{n!}{k!(n-k)!}$.

Centroid of a Triangle The intersection of the three medians in a triangle.

Chord A line segment connecting two points on the outside of a circle.

Circle A round shape consisting of points that all have the same distance (called the radius) from the center of the circle.

Circumcenter of a Triangle The intersection of the three perpendicular bisectors in a triangle. Also the center of the circle that circumscribes a triangle.

Circumference The perimeter of a circle.

Circumscribe To draw a shape outside another shape so that the boundaries touch.

Coefficient The number being multiplied by a variable or power of a variable. For example, the coefficient of x^3 in $5x^5 + 4x^3 + 2x$ is 4.

Complement In probability, the complement of a set is all elements outside the set.

Composite Number A number that is not prime.

Congruent Two shapes or figures that are exactly the same.

Cube A solid figure formed by 6 congruent squares that all meet at right angles.

Deck of Cards A standard deck of cards has 52 cards. There are 4 suits (clubs, diamonds, hearts, and spades) with each suit having cards of 13 ranks (A (ace), $2, 3, \ldots, 10, J$ (jack), Q (queen), and K (king)).

Degree of a Polynomial The highest power of a variable in the polynomial. For example, the degree of $2x^3 - 5x^6 + 2$ is 6.

Denominator The bottom number in a fraction.

Diagonal A line segment connecting two vertices of a shape or solid that is not an edge of the shape or solid.

Diameter A chord passing through the center of a circle. The diameter has length that is twice the radius.

Die or Dice A standard die (plural is dice) has 6 sides. Each of the 6 sides has the same chance when the die is rolled.

Digit One of $0, 1, 2, \ldots, 9$ used when writing a number.

Discriminant The expression $b^2 - 4ac$ for a quadratic equation $ax^2 + bx + c = 0$.

Distinguishable Objects Objects that are different.

Divisible A number is divisible by another number if there is no remainder when the first number is divided by the second. For example, 35 is divisible by 7.

Divisor A number that evenly divides another number. For example, 6 is a divisor of 48. Also called a factor.

Edge A line segment connecting two vertices on the outside of a shape or solid.

Equally Likely Having the same chance of occurring.

Equiangular Polygon A shape with all equal angles.

Equilateral Polygon A shape with all equal sides.

Equilateral Triangle A regular triangle, one with three equal sides and three equal angles.

Even Number A number divisible by 2.

Exponent The number another number is raised to for powers. For example, in a to the power of b (a^b), the exponent is b.

Exponential Form (of a complex number) A complex number written in the form $e^{i \cdot r \cdot theta}$ for real number r and angle θ.

Face The shape or polygon on the outside of a solid region.

Factor of a Number A number that evenly divides another number. For example, 6 is a factor of 48. Also called a divisor.

Factorial The symbol ! where $n! = n \times (n-1) \times (n-2) \cdots \times 1$.

Fraction An expression of a quotient. For example, $\frac{1}{2}$ or $\frac{9}{7}$.

Function A function is a rule that associates exactly one output with every input. Often described using an equation.

Geometric Sequence A sequence where the ratio between one term and the next is constant.

Greatest Common Divisor/Factor (GCD/GCF) The largest number that is a divisor/factor of two or more numbers.

Incenter of a Triangle The intersection of the three angle bisectors in a triangle. Also the center of a circle that is inscribed inside a triangle.

Indistinguishable Objects Objects that are the same.

Inscribe To draw a shape inside another shape so that the boundaries touch.

Intersecting Lines or curves that cross each other.

Intersection of Two Sets The set of objects that are in both of the two sets. Denoted using ∩. For example, $\{2,3\} \cap \{3,4,5\} = \{3\}$.

Isosceles Triangle A triangle with two equal sides and two equal angles.

Least Common Multiple (LCM) The smallest number that is a multiple of two or more numbers.

Mean The sum of the numbers in a list divided by the how many numbers occur in the list. Also called the average.

Median The number in the middle of a list when the list is arranged in increasing order.

Median of a Triangle A line connecting a vertex in a triangle to the midpoint of the opposite side.

Midpoint The point in the middle of a line segment.

Mode The number or numbers occurring most often in a list of numbers.

Multiple A number that is an integer times another number. For example, 72 is a multiple of 8.

Numerator The top number in a fraction.

Obtuse Angle An angle between $90°$ and $180°$.

Odd Number A number not divisible by 2.

Orthocenter of a Triangle The intersection of the three altitudes in a triangle.

Parallel Lines Lines that do not intersect.

Perfect Cube A number that is another number cubed. For example, $64 = 4^3$ is a perfect cube.

Perfect Square A number that is another number squared. For example, $64 = 8^2$ is a perfect square.

Perimeter The length/distance around the outside of a shape.

Perpendicular Bisector A line perpendicular to and passing through the midpoint of a line segment.

Pi (π) A number used often in geometry. $\pi = 3.1415926\ldots \approx 3.14 \approx \dfrac{22}{7}$.

Polar Form (of a complex number) A complex number written in the form $r\cos(\theta) + i \cdot r\sin(\theta)$ for real number r and angle θ.

Polygon A shape formed by connected line segments.

Polynomial A function that is made of adding multiples of powers of a variable. For example, $f(x) = x^4 + 3x^2 + 2x - 3$.

Prime Factorization The expression of a number as the product of all its prime factors. For example, 24 has prime factorization $2 \times 2 \times 2 \times 3 = 2^3 \times 3$.

Prime Number A number whose only factors are one and itself.

Proportional Ratios Ratios that have equal values when expressed in fraction form. For example, $2 : 3$ is proportional to $8 : 12$.

Quadratic A polynomial with degree 2. Often written in the form $ax^2 + bx + c$.

Quadrilateral A shape with four sides.

Quotient The integer quantity when dividing one number by another. For example, the quotient of $38 \div 5$ is 7 as $38 = 7 \times 5 + 3$.

Radius of a Circle The distance from the center of the circle to any point on the outside of the circle.

Randomly Chosen for a group of objects. Unless specified, the chance of choosing each object is the same as any other object.

Rank of a Card See Deck of Cards.

Ratio A relation depicting the relation between two quantities. For example $2 : 3$ or $\frac{2}{3}$ denotes that for every 3 of the second quantity there are 2 of the first quantity.

Rational Number A number that can be written as a fraction.

Reciprocal One divided by the number. For example, the reciprocal of 7 is $\frac{1}{7}$.

Rectangle A quadrilateral with four right angles (an equiangular quadrilateral).

Rectangular Form (of a complex number) A complex number written in the form $a + b \cdot i$ for real numbers a and b.

Regular Polygon A polygon with all equal sides and all equal angles (equilateral and equiangular).

Remainder The quantity left over when one integer is divided by another. For example, the remainder of $38 \div 5$ is 3 as $38 = 7 \times 5 + 3$.

Rhombus A quadrilateral with four equal sides (an equilateral quadrilateral).

Right Angle A 90° angle.

Right Triangle A triangle containing a right angle.

Root of a Function A value of x such that the function evaluates to zero. For example, $x = 2$ is a root of the function $f(x) = x^2 - 4$.

Sample Space In probability, the sample space is the set of all outcomes for an experiment.

Scalene Triangle A triangle with three unequal sides and three unequal angles.

Sector The region formed by an arc and the two radii connecting the ends of the arc to the center of the circle.

Sequence An ordered list of numbers.

Set An unordered collection or group of objects without repeated elements. Denoted using curly brackets. For example, $\{1, 2, 3, 4\}$ is the set containing the integers $1, \ldots, 4$.

Similar Shapes or solids that have the same angles and sides that share a common ratio.

Simplest Radical Form An expression containing a radical such that the number inside the radical is an integer that has no perfect squares.

Skew Line Lines in 3-D space that neither intersect nor are parallel.

Sphere A round solid consisting of points that all have the same distance (called the radius) from the center of the sphere.

Square A shape with four equal sides and four equal angles (a regular quadrilateral).

Subset A set of objects that is contained inside a larger set of objects. Denoted using $\subseteq$. For example $\{2,3\} \subseteq \{1,2,3,4\}$.

Suit of a Card See Deck of Cards.

Surface Area The total area of all the faces of a solid.

Tangent Line A line touching a shape or curve at exactly one point.

Trapezoid A quadrilateral with one pair of parallel sides.

Triangle A shape with three sides.

Union of Two Sets The set of objects that are in one or both of the two sets. Denoted using $\cup$. For example, $\{2,3\} \cup \{3,4,5\} = \{2,3,4,5\}$.

Venn Diagram A diagram with circles used to understand the relationship between overlapping sets.

Vertex The intersection of line segments, especially the intersection of sides or edges in a shape or solid.

Volume The amount of space a solid region takes up.

With Replacement When choosing objects with replacement, a chosen object is returned to the others allowing it to be chosen more than once.

3.3 ZIML Answers

ZIML October 2017 Varsity Division

Problem 1:	6	Problem 11:	49
Problem 2:	29	Problem 12:	4872
Problem 3:	78	Problem 13:	141
Problem 4:	2	Problem 14:	2016
Problem 5:	1033	Problem 15:	-1
Problem 6:	480	Problem 16:	1
Problem 7:	2	Problem 17:	2
Problem 8:	22	Problem 18:	153
Problem 9:	33	Problem 19:	3
Problem 10:	30	Problem 20:	0

ZIML November 2017 Varsity Division

Problem 1:   362880   Problem 11:   43

Problem 2:   27   Problem 12:   17550

Problem 3:   143   Problem 13:   122

Problem 4:   171   Problem 14:   0.5

Problem 5:   62.5   Problem 15:   7

Problem 6:   23430   Problem 16:   13

Problem 7:   2   Problem 17:   53

Problem 8:   46   Problem 18:   16

Problem 9:   630   Problem 19:   7535

Problem 10:   18   Problem 20:   4

ZIML December 2017 Varsity Division

Problem 1:	80	Problem 11:	7
Problem 2:	36	Problem 12:	121
Problem 3:	7	Problem 13:	-8
Problem 4:	256	Problem 14:	10
Problem 5:	10081	Problem 15:	31
Problem 6:	4	Problem 16:	128
Problem 7:	49	Problem 17:	148
Problem 8:	321	Problem 18:	49
Problem 9:	95	Problem 19:	37
Problem 10:	36	Problem 20:	-1

ZIML January 2018 Varsity Division

Problem 1:	24	Problem 11:	266
Problem 2:	25	Problem 12:	16
Problem 3:	10395	Problem 13:	10
Problem 4:	21	Problem 14:	7
Problem 5:	2000	Problem 15:	3
Problem 6:	12600	Problem 16:	155
Problem 7:	120	Problem 17:	10
Problem 8:	9	Problem 18:	-417
Problem 9:	6	Problem 19:	806
Problem 10:	14112	Problem 20:	1

ZIML February 2018 Varsity Division

Problem 1:	6300	Problem 11:	13
Problem 2:	900	Problem 12:	49
Problem 3:	11.7	Problem 13:	4035
Problem 4:	5	Problem 14:	130
Problem 5:	4100	Problem 15:	2
Problem 6:	42	Problem 16:	17
Problem 7:	18	Problem 17:	9
Problem 8:	169	Problem 18:	6
Problem 9:	3334	Problem 19:	4037
Problem 10:	6	Problem 20:	1223

ZIML March 2018 Varsity Division

Problem 1:	54	Problem 11:	45
Problem 2:	100	Problem 12:	5
Problem 3:	29	Problem 13:	4
Problem 4:	4351	Problem 14:	61.7
Problem 5:	18	Problem 15:	171
Problem 6:	10	Problem 16:	24
Problem 7:	46	Problem 17:	31
Problem 8:	1001	Problem 18:	486
Problem 9:	21	Problem 19:	997999
Problem 10:	120	Problem 20:	4

ZIML April 2018 Varsity Division

Problem 1:	7	Problem 11:	-0.2
Problem 2:	1008	Problem 12:	4
Problem 3:	31	Problem 13:	1
Problem 4:	8	Problem 14:	0.25
Problem 5:	103	Problem 15:	47
Problem 6:	15	Problem 16:	14
Problem 7:	25	Problem 17:	21
Problem 8:	218	Problem 18:	70
Problem 9:	112	Problem 19:	11
Problem 10:	2.4	Problem 20:	1000

ZIML May 2018 Varsity Division

Problem 1:	834	Problem 11:	105
Problem 2:	92	Problem 12:	26
Problem 3:	3.5	Problem 13:	2
Problem 4:	3	Problem 14:	10
Problem 5:	10073	Problem 15:	6.25
Problem 6:	1.57	Problem 16:	230
Problem 7:	86	Problem 17:	7
Problem 8:	143	Problem 18:	1681
Problem 9:	0.75	Problem 19:	434
Problem 10:	99.99	Problem 20:	866

ZIML June 2018 Varsity Division

Problem 1:	4.6	Problem 11:	42
Problem 2:	1	Problem 12:	10
Problem 3:	32.5	Problem 13:	16
Problem 4:	14851	Problem 14:	30
Problem 5:	12	Problem 15:	24
Problem 6:	5050	Problem 16:	3
Problem 7:	47	Problem 17:	4500
Problem 8:	11	Problem 18:	1.69
Problem 9:	528	Problem 19:	7
Problem 10:	1	Problem 20:	0.5

www.ingramcontent.com/pod-product-compliance
Lightning Source LLC
Chambersburg PA
CBHW072307210326
41519CB00057B/3054